PRACTICAL PHOTOMICROGRAPHY

PRACTICAL
PHOTOMICROGRAPHY

J. Bergner, E. Gelbke and W. Mehliss

THE FOCAL PRESS

LONDON and NEW YORK

Translated by K. S. Ankersmit

German edition

Einführung in die praktische Mikrofotografie
© Fotokinoverlag Halle

Printed in East Germany
210/156/66

FOREWORD

Photomicrography nowadays has a wide application in most spheres of technology and science. The progress made in the design of apparatus and accessories has contributed considerably to the documentary value of the photomicrograph. However, the results are only of value when any possible distortions, introduced by the operator, are excluded. This entails a perfect knowledge of micrographic instruments.

There are a number of books on the techniques used in photomicrography, as well as excellent instruction manuals supplied by manufacturers. All this literature, however, presumes a certain level of technical, optical and photographic knowledge without which it is impossible to obtain optimum results. We have learnt through personal experience that students in subjects which require microscopy and photomicrography as working tools, only acquire the general information necessary for their operation, without studying the reasons for the manipulation of optical instruments, which are so necessary to achieve their fullest possible use. It is therefore one of the main objects of this book to supply the essential optical knowledge of the photomicrographic instrument.

Beyond this, the present book aspires to become a frequently consulted source of information during practical work. For this reason, technical information on practical photomicrography takes priority, while all major specialized techniques are included. Tables, graphs and practical advice should be of considerable assistance in the laboratory. A bibliography of general and specialized subjects is included as an incentive for further study, and supplies more detailed information on those points which are only dealt with as accepted facts in this book.

It was finally decided to omit nomograms, since such charts, although having certain demonstration value in a book, are nevertheless of very little practical use. Furthermore, photomicrography is hardly a subject for nomograms, because nomographic aids in the shape of exposure and magnification calculators have not proved entirely successful.

It was also decided not to elaborate the purely photographic side of the subject in this book, as it may be presumed that the reader has a working knowledge of the processing of photographic materials. Only questions directly arising from photomicrographic practice have been discussed.

We wish to express our gratitude to all colleagues who have assisted us with advice and practical help.

CONTENTS

INTRODUCTION

Photomicrography and macrophotography both produce photographs larger than the original. Usually, the term macrophotography is taken to describe magnifications from same-size to about 10 diameters, while photomicrography takes over from there. Optical methods enable us to discern smaller details than is possible with the unaided eye. One of the tasks of macrophotography and photomicrography is to produce a photographic record of maximum fidelity of the object observed through a photographic lens (magnifying glass in its most primitive form) or microscope.

By using invisible radiation, the scope of the work can be extended to yield important information on the structure, texture, nature or composition of the material under examination. The methods used in photomicrography for producing records of both mounted and unmounted specimens are of a physical nature. Fig. 1 shows the versatility of the methods of reproduction. These photomicrographs show seven different ways of photographing a single specimen. Each record is valid and shows certain properties and aspects of the specimen. Which technique will be chosen therefore depends on the desired type of record. The photomicrographs must be taken so as to apply physical (here mainly optical) laws, depending on the need to reveal absorption, surface, ultramicroscopic or molecular structure, or the structure of crystals.

We shall begin the book with the optical principles needed for rational working with the microscope and photomicrographic equipment.

1 a   1 b

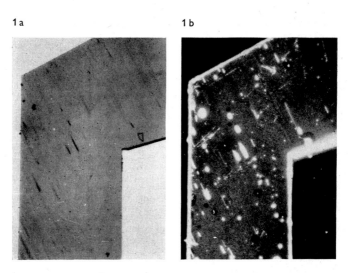

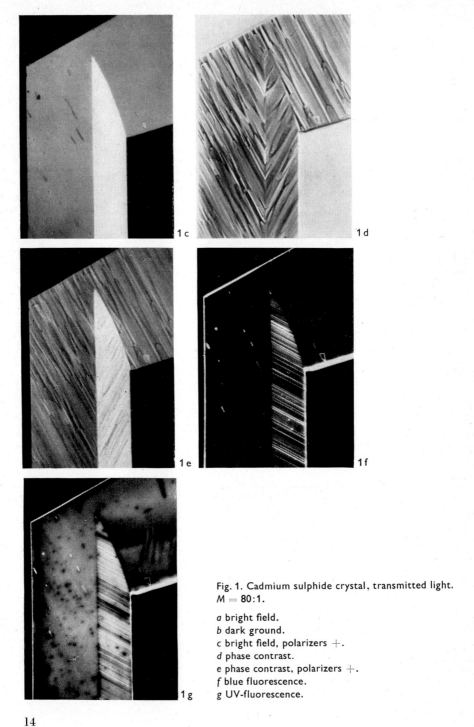

Fig. 1. Cadmium sulphide crystal, transmitted light.
$M = 80:1$.

a bright field.
b dark ground.
c bright field, polarizers $+$.
d phase contrast.
e phase contrast, polarizers $+$.
f blue fluorescence.
g UV-fluorescence.

I. GENERAL OPTICAL PRINCIPLES

1.1 Lenses

Photomicrographic equipment consists basically of lens arrangements. We shall therefore consider first the optical principles. Strictly speaking, the following basic principles are only valid for narrow parallel beams and light of a single given wavelength (monochromatic light). As soon as a wider beam or 'white' light, consisting of a multitude of wavelengths, is used, aberrations will occur which can be eliminated by suitable lens shapes and lens combinations.

1.11 Converging (convex or positive) lenses

1. The optical axis of a lens is the line joining the centres of curvature C_1 and C_2 of the two surfaces (Fig. 2).

2. The field in front of the lens is the object space, behind the lens the image space.

3. P_1 and P_2 are the points of intersection of the optical axis with the lens surfaces and are known as the poles.

4. The rear principal focus F_2 is the point on the optical axis on which rays travelling parallel to the axis converge after having passed through the lens.

5. Rays travelling from the front principal focus F_1 will emerge from the lens parallel to the lens axis.

6. The planes at right angles to F_1 and F_2 are called focal planes, and the rays passing through these points are called focal rays.

7. By extending the focal rays to points within the thickness of the lens where they meet their corresponding rays emerging from the lens parallel to the optical axis, we obtain two points on the principal planes of the lens, which are at right angles to the optical axis through these points.

Diagrams often show these two principal planes instead of the lens itself. The separation between these two planes depends on the thickness of the lens and the refractive index of the glass, while their distance from the centres of curvature of the lens depends on its shape. When it is possible to ignore the lens thickness, a diagram will show the single principal plane of a 'thin' lens.

8. The points of intersection of the optical axis with the two principal planes are called the front and rear nodal points (N_1 and N_2 respectively). These are also called 'principal points'.

9. The distances between the focal points and the corresponding nodal points are called front focal length f_1 and rear focal length f_2. With lenses and lens combinations, these focal lengths are always in opposite directions, but will be of equal length when the optical medium is the same in front and behind the lens. The symbol F

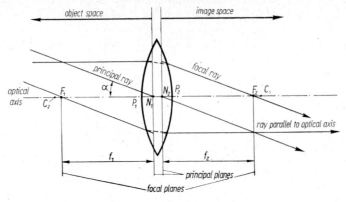

Fig. 2.

N_1 = front nodal point.	u = distance from the object to the lens.
N_2 = rear nodal point.	
F_1 = front principal focus.	v = distance from the image to the lens.
F_2 = rear principal focus.	
P_1 = pole of front lens surface.	F_1, F_2 = focal length of the lens.
P_2 = pole of rear lens surface.	

is normally used to designate the focal length of a lens, irrespective of the position of the focal points.

10. Rays going through the nodal points are known as the principal rays. Conjugate principal rays are always parallel. α is the angle subtended to the optical axis by the principal ray.

11. Lens measurements are determined from the nodal points. Lengths emanating from N_2 in the direction of travel receive a positive sign, those emanating from N_1 opposite to the direction of travel receive a negative sign. According to the direction of their rear principal focus, converging lenses are called positive lenses.

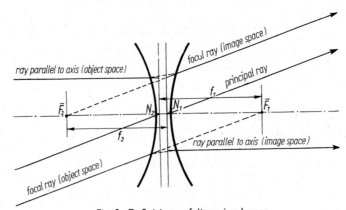

Fig. 3. Definitions of diverging lenses.

1.12 Diverging (concave or negative) lenses. Diverging lenses spread the incident beam of light (Fig. 3). Rays travelling parallel to the optical axis emerge from the lens as if they had originated from a point at a finite distance in front of the lens. This point can be determined by continuing the rays that emerge from the lens. Agreeing with the terminology used for converging lenses, this point is called the rear focal point F_2. As this point is merely apparent, it is given the term 'virtual'. This corresponds with an equally virtual front focal point F_1 behind the lens. Rays travelling to the front focal point emerge from the lens parallel to the optical axis. As F_2 is negative, the lens is thus called a negative lens.

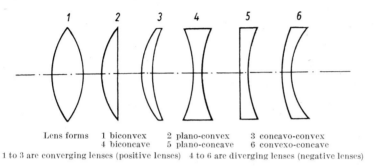

Lens forms 1 biconvex 2 plano-convex 3 concavo-convex
4 biconcave 5 plano-concave 6 convexo-concave
1 to 3 are converging lenses (positive lenses) 4 to 6 are diverging lenses (negative lenses)

Fig. 4. Lens shapes.

1.13 Refraction. The degree of the converging or diverging action of a lens, i.e. its refractive power, is inversely proportional to its focal length. Refractive power is expressed in diopters, the reciprocal of the focal length expressed in metres.

1.14 Lens types. In order to eliminate optical aberrations, lenses are given different shapes, as shown in Fig. 4.

1.2 Image formation by lenses

1.21 Image formation by converging lenses. Rays parallel to the optical axis, reaching the lens from a very distant object ('infinity') are converged on the principal focus or focal point of the lens, i.e. a point on a very distant object forms a real image at the focal point. As either the lens or the direction of the light beam can be reversed, it is immaterial which of the two points is considered as object, and which as image. In other words, both bear a definite relation to each other: they are conjugates.

Fig. 5 shows how an image y' is formed from an object y normal to the optical axis, by tracing the path of light from O_1 and O_2 to O_1' and O_2' with the help of a parallel, focal and principal ray. The graphical construction shows that the image plane of an object plane normal to the optical axis, will also be normal to this axis (O-O'). It follows that any given object plane will correspond to a certain image plane. If an object y of a given dimension approaches the lens (the five positions in

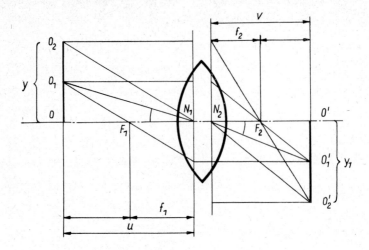

Fig. 5. Image formation by converging lens.

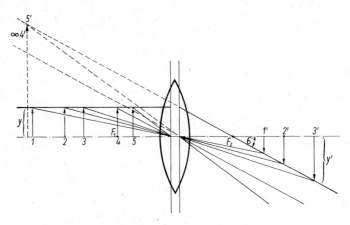

Fig. 6. Image formation of a real object by a converging lens.

Fig. 6), the image y' will occupy positions 1'-5'. As long as the object distance is greater than the focal length of the lens, y' will be real and its size will become greater as it is taken farther away from the lens. The largest size is apparently reached when it is at infinity, with the object in the focal plane (position 4). When y comes still closer to the lens (position 5) the rays travelling from the individual points of the object will be deflected by the lens, and a virtual image will be formed in the backward continuation of these rays. This virtual image will become increasingly smaller as the object comes closer to the lens, and will be apparently reduced to the size of the object when it has reached the rear principal plane and the object is placed in the front principal plane.

18

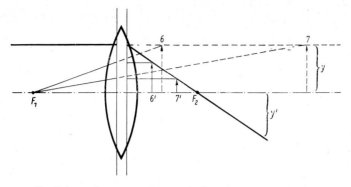

Fig. 7. Image formation of a virtual object by a converging lens.

An image can also be formed within the rear focal length when the object is situated in the image space. This is possible when the object presented to the lens is not a real object, but an image formed by another lens, independent of the lens under consideration. Fig. 7 shows such a 'virtual object', while Table 1 gives a summary of the ratios of reproduction.

1.22 Image formation by diverging lenses. The graphical construction is exactly the same as in the case of converging lenses. The different position of the focal planes, however, is to be considered. For the sake of clarity, Fig. 8 shows a 'thin' lens. The ratios of reproduction are given in Table 2. In photomicrography, the most frequently occurring case will be the formation of a real image (Fig. 8, *2′*).

Finally, let us draw attention to something which is valid for image formation with both types of lenses. When the height of the object remains constant during its approach to the lens (i.e., when the highest point of the object remains on the same ray parallel to the optical axis), the highest point of the image (which will gradually increase in size) will remain on the same focal ray. The angle σ under which the image is seen from F_2, is constant.

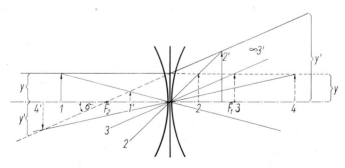

Fig. 8. Image formation by a diverging lens.

1.3 Scale of reproduction and the principal lens equations

The scale of reproduction (M) plays an important part in photomicrography. It is the ratio between the linear dimension of image y' and that of object y:

$$M = y'/y \qquad (1)$$

It is usual to express M as the ratio to object dimension $= 1$ (e.g. $M = 25:1$). A few more definitions are necessary to express image formation in equations (see Fig. 5):

$u = $ object distance $= N_1 O \qquad U = \dfrac{1}{u} = $ reciprocal of object distance

$v = $ image distance $= N_2 O' \qquad V = \dfrac{1}{v} = $ reciprocal of image distance

We should also remember the plus and minus signs mentioned under point 12 in 1.11. The following equations can now be derived from Fig. 5:

$$(u - f_1) \cdot (v - f_2) = f_1 \cdot f_2$$

or

$$(u - f_1) \cdot (v - f_2) = -f_2^2 \qquad (2)$$

$$\frac{1}{v} = \frac{1}{u} + \frac{1}{f_2} \quad \text{or} \quad V = U + D \qquad (3)$$

(the values for V, U and D are expressed in diopters)

$$M = (-)\frac{f_1}{u - f_1} \qquad\qquad M = (-)\frac{v - f_2}{f_2}$$

$$f_1 = (-)(u - f_1) \cdot M$$

$$f_2 = (-)\frac{v - f_2}{M}$$

$$u - f_1 = (-)\frac{f_1}{M} \qquad (4)$$

$$v - f_2 = (-)f_2 \cdot M \qquad (5)$$

$$u = (-) f_1 \cdot \left(\frac{1}{M} + 1\right) \qquad (6)$$

$$v = (-) f_2 \cdot (M + 1) \qquad (7)$$

When using equation (3) it is necessary to take into account the direction of the ray-lengths by correct use of the plus or minus sign. The minus sign shown in brackets in equations (4)—(7) is always operative when the values have been inserted with their correct signs, and when it is important to determine the direction. This makes M negative in the example in the diagram, which indicates that the image is inverted.

1.4 Scale of reproduction and magnification

The scale of reproduction is the relationship between the size of the object and its image. To determine the scale of reproduction it is necessary to compare the sizes of object and image either directly, or by means of a scale situated at these positions,

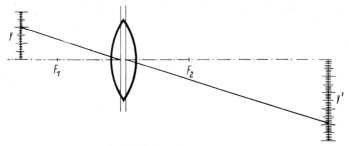

Fig. 9a. Scale of reproduction.

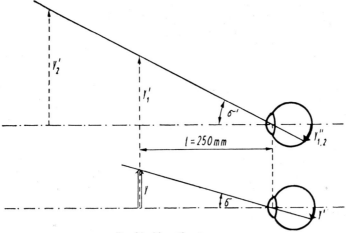

Fig. 9b. Magnification.

such as a micrometer slide, engraved with 100 or more divisions per millimetre (Fig. 9a).

On the other hand, magnification refers to the ratio of the virtual image size to the object size, i.e. angle σ' and σ, under which image y' and object y appear to the observer (Fig. 9b). With certain limitations, the magnification indicates how many times the image on the retina of the observer's eye is larger when looking at y' than when looking at y.

Magnification and scale of reproduction are only numerically equal when object and image are situated at the same distance from the eye (y_1' and y in Fig. 9b). In order to determine the magnification it is important to choose the best possible distance at which the image is to be observed by the eye. This has been found to be 250 mm $= l$ and is termed 'minimum distance for comfortable vision'.

The concept and term 'magnification' should therefore only be used for visual observation. We can define magnification as follows:

$$Mg = \frac{\sigma'}{\sigma} \tag{8}$$

21

Magnification is usually expressed in diameters, i.e. a certain number of 'times' (linear), e.g. ×250.

The following example illustrates how necessary it is to keep the concepts of scale of reproduction and magnification separate. A photomicrograph has been made so that it shows fine detail at a viewing distance of 250 mm. In this case, magnification equals scale of reproduction. When projecting the photomicrograph on a screen, to enable a number of people to observe the picture from a greater distance of e mm, the picture will be considerably enlarged. The angle of vision σ, at which the photomicrograph was observed at the distance of 250 mm must be maintained (see Fig. 9b). The secondary enlargement factor will be:

$$p = \frac{e}{250}$$

so that $M = p \cdot Mg$ in the present example. The object is reproduced on the photomicrograph at a scale of reproduction M, but is seen during projection at a magnification Mg. As p represents four times the observation distance measured in metres, the difference between M and Mg can be quite considerable.

1.5 Field angle limitation and aberrations

1.51 Introduction to the magnifying lens. The magnifier is used to increase the angle of vision for the observation of small objects which are very close to the eye. In its simplest form, the magnifier consists of a single converging lens. An observer with normal vision will place the object at a distance equal to the focal length of the magnifying lens, thus being able to look at it with relaxed eye accommodation. Fig. 10 shows the

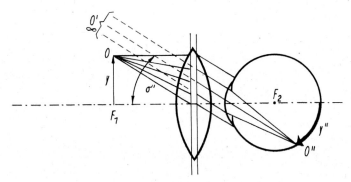

Fig. 10. Magnifier and the eye.

graphical construction in conjunction with the eye. Rays emanating from point O and contributing to the image formation on the retina, reach the eye at an angle σ' subtended to the optical axis of the magnifying lens. The eye will see the object y at the same

22

angle at infinty. When no magnifier is used, the image will be observed by the eye at a distance of 250 mm at an angle σ (lower part of Fig. 9 b). The magnification is therefore:

$$Mg_{\text{magnifier}} = \frac{\tan \sigma'}{\tan \sigma} = \frac{y}{f_2} \cdot \frac{250}{y}$$

$$= \frac{250}{f_2} = \frac{D}{4} \tag{9}$$

D is expressed in diopters. In the present case, magnification is independent of the distance between magnifier and eye. This, as for all other observations so far, is only valid for a narrow beam of light. The diameter of the light beam which can reach the eye when using a magnifier is restricted by the size of the eye's pupil. Its size and location govern the aperture, and influence the position, of the cone of light rays in the object space used for observation. When looking at points away from the optical axis, the eye is turned: the pupil moves away from its original position and parts of the optical path of the rays will now be used, which pass through the magnifying lens more or less off-centre. By increasing the distance between the magnifier and the eye, the object area which can be observed, is decreased, since the field is limited by the lens mount.

When carrying out this experiment with a simple uncorrected magnifying lens of average power, perhaps with a biconvex lens of a focal length of 50 mm, and selecting as object a simple pattern with parallel straight lines or linearly arranged dots (e.g. a piece of millimetre paper or a coarse printing screen) the following conclusions will be arrived at:

1. At the shortest possible distance between lens and eye, the central part of the object is correctly rendered. Looking towards the field edge, lines will be seen which do not go through the centre of the field and show barrel or pin-cushion distortion. It is also impossible to see parts of the object near the edge of the field which are entirely sharp. Moreover, these parts will show coloured fringes. Their image will be improved by shortening the distance between lens and object.

2. By increasing the distance between eye and lens, one will observe an additional marked decrease in the diameter of the useful field—apart from the above-mentioned limitation of the field by the lens mount.

These experiments show that the pencils of rays coming from the various object parts, and passing through both the lens and the eye pupil, have very different qualities from those of their image-forming characteristics. The influence of the path-of-rays limitation thus observed is basically valid for all optical instruments.

1.52 Field of view limitation. To understand the concept of field of view limitation, it will help first to study the behaviour of an optical system consisting of two lenses separated by a diaphragm. We should remember that the mounts of the two lenses also act as diaphragms. When the observer looks through this system from the axis point of the object, he will see the diaphragms and their images under different angles. The diaphragm, or its image, which he will see under the smallest angle, is called the entrance pupil (EP) of the system (Fig. 11). It is the basis of a cone of rays which enters the optical system from the object axis. The angle σ between the optical axis and the ray travelling to the edge of the diaphragm is half the aperture angle in the object field. In the system constituting lens plus eye, the image of the eye pupil formed by the lens acts as entrance pupil.

Corresponding to the EP, the exit pupil (AP) and the half aperture angle σ' will appear in the image field of the system. The mechanical diaphragm, which is the basis of the two pupils, is called aperture diaphragm (AD). According to its position, it can be identical with one or the other of the two pupils. The pupils are therefore either images of the aperture diaphragm, or one of the pupils is the image of the other.

Apart from the effect of the aperture diaphragm, the path of rays of an optical system is also usually limited at another place, namely in relation to the dimension of the field which is being imaged. In the case of the magnifying lens, this is done— although imperfectly—by the lens mount. The limitation would be perfect in this case if a suitable diaphragm were arranged in the object plane. This would be called the field diaphragm (FD). An example of this is the transparency viewer. The field diaphragm may be placed in any plane conjugate to the object plane.

When yet other diaphragms are present in the path of rays—apart from aperture and field diaphragms—as for example the lens mount already mentioned, they are called 'shading diaphragms', and the images formed by them through parts of the system are called 'windows'. These diaphragms shade off the image away from the optical axis towards the margin of the field of vision; in other words, 'vignetting' takes place. The more off-axis the image points are situated, the more the lens mounts will move in front of the aperture diaphragm, with the result that these image points will receive less light. This is illustrated in Fig. 11, at the object points above point O_2.

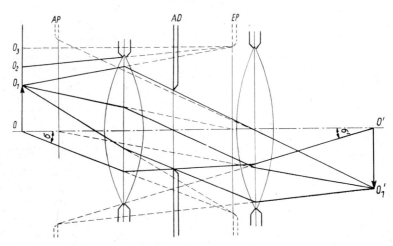

Fig. 11. Field of view limitation.

From this point upwards, the upper rim of the entrance pupil will be obscured by the lens mount. The beam of rays which the entrance pupil can accept is therefore vignetted.

1.53 Lens aberrations. As mentioned in section 1.1, the equations given are valid for simple lenses in a narrow beam of light. With wider beams of light, it is generally not possible to expect all the rays passing through the lens to come to a common focus. More or less pronounced deviations will occur, depending on the angle of incidence and shape of the lens, in the intersection of the individual rays with each other and with the optical axis. We shall only discuss here such aberrations as occur in the microscope image, and which can be influenced, at least partly, by the user.

The lens designer has quite an array of means for the correction of these aberrations, in the form of refractive power of different glass types, the combination of lenses of the same or different glasses, varying thicknesses and distances within an optical system, and the insertion of suitably sized diaphragms at the most appropriate places. This is not to say that all aberrations can be completely eliminated. They are, however, corrected to such a degree that they will not be objectionable when the optical system is used in the prescribed way—very important in the production of a photomicrograph.

1.531 Direct errors (spherical aberration). When a wide beam of rays parallel to the optical axis passes through a simple converging lens, rays passing through different concentric zones will come to a focus at different points of the optical axis (Fig. 12).

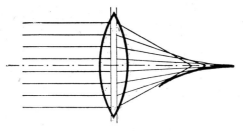

Fig. 12. Spherical aberration.

The greater the distance of such an annular zone from the optical axis, the shorter will be its focal length. Instead of a single focal point at the apex of a uniformly converging cone of rays, each zone will produce its own focal point at the apex of its own cone.

Image points, therefore, are not sharply defined in a single plane, since the image extends along the optical axis and is surrounded by an unsharp fringe at any given point along the axis.

Notwithstanding the high degree of spherical correction of camera lenses and microscope objectives, it is nevertheless often possible to observe the effect of spherical aberration, particularly when the cover glass used is not of the exact thickness for which the optical system has been computed. In this case, the image will be flat and without contrast, showing less overall sharpness (Fig. 13).

It is not sufficient to correct spherical aberration for a pair of points on the optical axis, since this would not guarantee the same correction for a plane element lying close to the optical axis. This demands the additional requirement that such an element will be rendered at the same scale of reproduction by the various concentric zones of the lens (aplanatic image formation). This can only be achieved with a given system for a single pair of points on the axis, the so-called 'aplanatic points'. At other points on the axis, more or less pronounced decreases in image quality will occur. When spherical aberration has been corrected in the way outlined above, the lens system is 'aplanatically corrected'.

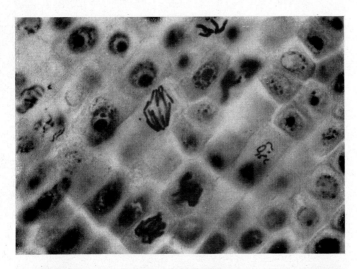

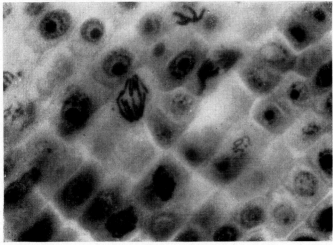

Fig. 13. Effect of spherical mitosis aberration. Brightfield—Transmitted light.
$M = 630:1$.

a correct thickness of cover glass.

b incorrect thickness of cover glass.

1.532 Oblique errors (asymmetric aberrations). An imaginary section through an off-axis pencil of rays, including both the axis of the pencil of rays and that of the lens (Fig. 14 in the plane of the drawing), will yield a pencil of rays of maximum eccentricity, defined by angle σ_B. In a section normal to the former, the axis will however coincide in the projection. Therefore this pencil of rays does not show asymmetry. Both pencils of rays are differently influenced by the lens.

1.5321 Astigmatism. The pencil of rays depicted in Fig. 14 is meant to represent an oblique, narrow pencil of parallel rays. After passing through the lens, the rays will converge in the usual way, but in the case of an uncorrected lens, the degree of convergence of the various sections through the pencil of rays is no longer constant. Of the two pencils of rays, chosen at random, the first-mentioned has the strongest, the other the weakest of the prevalent convergences. Instead of a single focal point (image point), the oblique pencil of rays produces two focal (or image) 'lines' in different planes and directions. For a given lens, their distance and length will be greater when the angle of incidence (obliqueness) is greater. Thus this pencil of rays which produces two image lines instead of an image point, is called 'astigmatic'. The image lines are situated in two astigmatic 'saucers' (Fig. 15,1) which touch each other in the centre. In microscopy, astigmatism can sometimes be observed at point-form objects near the margin of the image field.

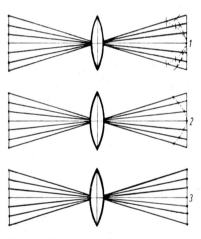

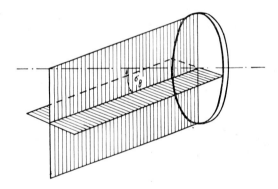

Fig. 15. Astigmatism and field curvature (left).

1 astigmatic "saucers".

2 field curvature.

3 correct image field.

Fig. 14. Oblique pencil of rays (right).

1.5322 Curvature of field. Astigmatism is eliminated when the two astigmatic image saucers are combined into a single whole (Fig. 15,2). The system will then show curvature of field, an aberration which only recently has been corrected in microscope objectives. Curvature of field makes it impossible to obtain a sharp image over the entire field; one can only obtain a single concentric zone.

27

1.5323 Distortion. A distortion of the image will occur towards the margin of the field as has been shown when the simple magnifying lens was discussed (Fig. 16). This distortion depends on the position of the aperture diaphragm owing to the fact that the scale of reproduction varies from the centre of the field towards the margin.

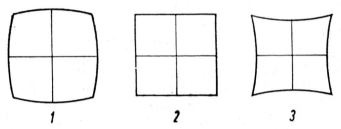

Fig. 16. Distortion. *1* barrel distortion. *2* correct. *3* pin-cushion distortion.

Pin-cushion or barrel distortion will result, according to whether the scale of reproduction increases or decreases towards the edges. In microscopes, this distortion is normally so small that it will only be observed in certain objects. In photomicrography, distortion can be corrected still further by a suitable combination of optical systems.

1.533 Chromatic errors

1.5331 Chromatic aberration. Chromatic aberration is a result of the refractive index of optical media being greater for shorter wavelengths than for longer wavelengths. This means that a simple lens has a shorter focal length for blue light than for red (Fig. 17).

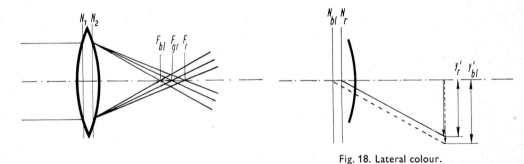

Fig. 17. Chromatic aberration.

F_{blue}, F_{green}, F_{red} focal points for blue, green and red light.

Fig. 18. Lateral colour.

N_{blue}, y'_{blue} principal plane and image for blue light.

N_{red}, y'_{red} principal plane and image for red light.

1.5332 Lateral colour (chromatic difference of magnification). This aberration occurs when the focal points of various colours coincide, but not their nodal points. Not only do the focal lengths differ, but also the scale of reproduction of individual colours for a given focal length (Fig. 18). Many microscope objectives are affected by this error, but it can be rendered unobjectionable (see 3.32 and 3.33).

1.6 The microscope

1.61 Image relationships. With the microscope, the magnified image formed by a positive system is not observed directly, as in the case of the magnifying lens, but by means of a second optical system. Fig. 19 shows the path of rays in its simplest form. The first lens system (the microscope objective) produces a real image of the object at a finite distance Δ behind its focal point. This distance constitutes the 'optical tube length', which is calculated for the objective. The image is now observed through a second lens system (the eyepiece) in the same way as a magnifying lens would be

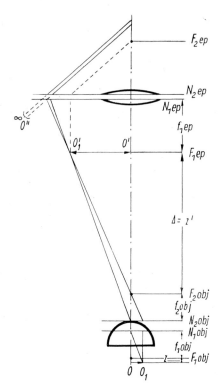

Fig. 19. Image formation by a two-stage microscope.

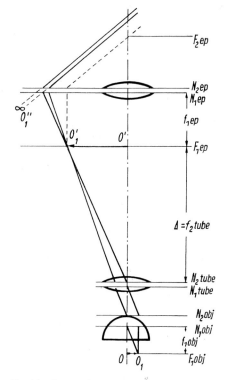

Fig. 20. Image formation by a three-stage microscope.

29

used. When the proper adjustment has been made for relaxed observation by an observer with normal vision, the real intermediate image produced by the objective will lie in the focal plane of the eyepiece and be seen at infinity.

We now have the following equations. According to (5), the scale of reproduction of the intermediate image is:

$$M_{obj} = \frac{-\varDelta}{F_{obj}} \tag{10}$$

and the magnification of the eyepiece according to (9):

$$Mg_{ep} = \frac{l}{F_{ep}} \tag{11}$$

This gives the following magnification of the microscope with two lens systems:

$$Mg_{mi} = M_{obj} \cdot Mg_{ep}$$
$$= \frac{-\varDelta \cdot l}{F_{obj} \cdot F_{ep}} \tag{12}$$

The minus sign indicates that the microscope image shows the inverted object, when no special means are used to keep it upright*.

Sometimes the image produced by the objective can lie at infinity, in the same way as that produced by a magnifying lens. In such cases, the image is observed by a telescope consisting of a special objective and an eyepiece (Fig. 20). The rays emanating from an object point emerge from the objective as a pencil of parallel rays, which is then converged to a real focal point in the focal plane of the second telescope objective (the tube system). The focal plane of the tube system is at the same time the object focal plane of the eyepiece which now acts as a magnifying lens. While the 'two-stage' image formation shown in Fig. 19 directly transforms the diverging pencil of rays, coming from the object point, into a converging pencil of rays by means of the microscope objective, the latter produces a pencil of parallel rays in the 'three-stage' image formation shown in Fig. 20, which is made convergent by

---

* When focused correctly, the object will be imaged at infinity by the microscope. The object is therefore placed in the object focal plane of the total system. The image focal plane lies in the exit pupil of the microscope, since rays entering the microscope which are parallel to the optical axis, will intersect in the image focal plane. According to (9) the focal length of the entire system is:

$$F_{mi} = \frac{l}{Mg_{mi}}$$

from which can be derived:

$$F_{mi} = \frac{F_{obj} \cdot F_{ep}}{-\varDelta} \tag{13}$$

$$F_{1,2} = \frac{F_1 \cdot F_2}{-\varDelta} \tag{13a}$$

The latter is the equation for the focal length of a system constituting two members, where $\varDelta$ indicates the distance between the two focal points facing each other.

30

means of the telescope·lens. According to (24, p. 49), the scale of reproduction of the combination microscope-objective plus the tube system is:

$$M_{\mathrm{obj}+\mathrm{tu}} = \frac{F_{\mathrm{tu}}}{F_{\mathrm{obj}}}$$

The task of the eyepiece is the same as before. The magnification of the three-stage microscope is therefore:

$$Mg_{\mathrm{mi}_3} = \frac{F_{\mathrm{tu}}}{F_{\mathrm{obj}}} \cdot \frac{1}{F_{\mathrm{ep}}} = \frac{l}{F_{\mathrm{obj}}} \cdot \frac{F_{\mathrm{tu}}}{F_{\mathrm{ep}}}$$

The first fraction represents, after transposition, the magnifying lens magnification of the objective, the second represents the magnification of the subsequent telescope. When extending the second equation by l, the total magnification can also be expressed as the product of magnifying lens magnification of the objective, the magnifying lens magnification of the eyepiece, and a factor which corresponds to the reciprocal of the magnifying lens magnification of the tube system. Therefore:

$$Mg_{\mathrm{mi}_3} = Mg_{\mathrm{obj}} \cdot Mg_{\mathrm{ep}} \cdot \frac{1}{Mg_{\mathrm{tu}}} = Mg_{\mathrm{obj}} \cdot Mg_{\mathrm{ep}} \cdot \frac{F_{\mathrm{tu}}}{l}. \qquad (14)$$

It is usual to find the data which most interest the user: M and Mg, engraved on the respective mounts. The quotient F_{tu}/l (when not unity) is indicated by the 'x' sign as a factor on the equipment. The same is valid for all other factors which have a bearing on the final magnification or scale of reproduction, as far as they are defined by the design of the apparatus. The separation of an objective designed for a given finite tube length in the combination of objectives for infinity tube and tube system, offers various technical advantages. The distance between the two members can be chosen within very wide limits, while it is easy to insert auxiliary equipment, such as reflection devices for incident light illumination (plane-parallel glass), which otherwise in a conical path of rays could easily lead to a loss in image quality. Moreover, the focal length of the tube lens, which in practice only represents $\varDelta$ in (10) and (12), can be adapted to the requirements of the entire system without much difficulty.

For the sake of simplicity, the following general considerations refer mainly to the two-stage microscope.

1.62 Image limitation

1.621 Image limitation with luminous specimens. Image limitation is obtained by the usual diaphragms and diaphragm images (Fig. 21). The exit pupil of the objective lies ideally in its rearmost focal plane (the position of the aperture diaphragm depends on the design of the system). With such an arrangement, the entry pupil lies virtually at infinity. The rays emanating from the various object points to the edges of the exit pupil thus travel under identical angles with the optical axis, or to the standard ray of the pencil of rays travelling from this point to the objective. As long as no vignetting by mechanical diaphragms occurs, quantity and direction of the light which travels from the various object points to the objective, will be the same, at equal radiation intensity. The aperture angle 2σ is therefore a measure of the quantity of radiation emanating from an object point, which the objective is maximally capable of receiving. *Abbe* has taken as its value the sine of half the aperture angle, multiplied by the refractive index of the medium of least optical density which is between the object and the objective. He called this the numerical aperture (N.A.):

$$\text{N.A.} = n \cdot sin\sigma \qquad (15)$$

The numerical aperture represents a performance number which could be compared to the relative aperture of a camera lens. In both systems, the quantity of light which the objective or lens is capable of receiving from the object, increases with the square of the performance numbers. We shall come back to the particular importance of the numerical aperture in microscopy.

The exit pupil of the compound microscope is the image of the objective's exit pupil produced by the eyepiece. When using the microscope, this pupil can be seen in the

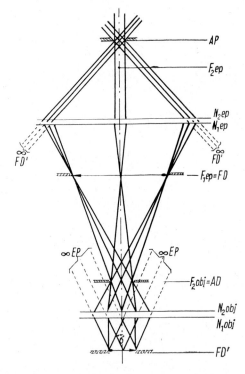

Fig. 21. Field limitation in the microscope.

shape of a small luminous disc behind the eyepiece (Ramsden circle). By visual observation, the eye's pupil is brought into their position. Their diameter can be given as:

$$2h' = 500\,\frac{N.A.}{Mg_{\mathrm{mi}}}\,\mathrm{mm} \tag{16}$$

The field diaphragm lies in the object focal plane of the eyepiece, i.e. at the place of the intermediate image of the object produced by the objective, and is therefore a conjugate of the object and image planes.

1.622 Image limitation with illuminated specimens. Most microscope specimens are illuminated, and not luminous objects. Most specimens are transparent, requiring transmitted illumination. Modern microscopes are therefore equipped with a condenser or condenser system, which directs the light through the specimen and gives the best possible uniform illumination compatible with the full exploitation of the light source in use. Let us assume that the axis of the illumination cone coincides with the optical axis of the microscope, and that its aperture angle 2σ is adjustable. To obtain this, a diaphragm is arranged in the focal plane of the condenser. To simplify matters, let us also assume that a ground-glass or opal glass is arranged in its immediate vicinity, so that a uniform illumination of the diaphragm is ensured. The various points of the image plane will then receive light from the image of the illuminated diaphragm area, which is at infinity, the exit pupil of the illumination system (AP_{con}). Under the arrangement described above, the aperture cones of the illumination rays aiming at the various image points, will be of equal size and travel in the same direction, namely, parallel to the axis (Fig. 22). The diaphragm mentioned is called the 'aperture diaphragm of the illumination device', or simply 'condenser diaphragm'. In this assumed position, the condenser diaphragm is a conjugate of the image focal plane of the objective, and hence of the exit pupil of the entire microscope. If the diaphragm is closed to such an extent that its image does not entirely fill the exit pupil of the objective (as is usual in most cases), then the size of the illuminated exit pupil of the microscope depends on the size of the diaphragm. In other words, the latter becomes the aperture diaphragm of the entire microscope.

In practice, however, it is often necessary to deviate from the ideal arrangement as shown in Fig. 22, since it is not always possible to arrange the aperture diaphragm in the immediate proximity of the focal plane of the condenser. When working with medium and high-power objectives, these deviations will have no importance, since the image of the aperture diaphragm of the condenser will lie close enough to the exit pupil of the objective. The microscopic image is not perceptibly impaired. The deviations between the two pupil positions, however, become more distinct when low-power objectives are used, because the focal length of the condenser will be increased by removing its top component (see 3.221). In this case, the exit pupil of the condenser will lie at a finite distance from the condenser, i.e. in the direction of the light source. The pencils of rays travelling from a given point of the aperture diaphragm will now leave the condenser as diverging pencils of rays (Fig. 23). The axes of the light cones which illuminate oblique object points, subtend a larger or smaller angle to the optical axis; these points receive oblique illumination. The attentive observer can see this oblique illumination by comparison with similar details in the centre and near the edge of the object field (see 2.1). The parallax occurring between the exit pupil of the objective and the image of the condenser aperture

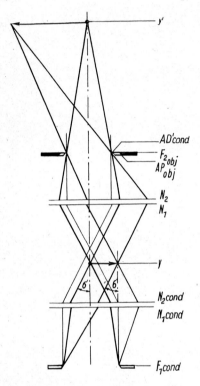

Fig. 22. Path of light between condenser diaphragm and intermediary image of object (ideal case).

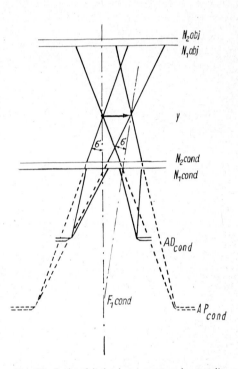

Fig. 23. Path of light between condenser diaphragm and object (exit pupil outside the focal length of the condenser).

diaphragm can be observed by looking into the tube without the eyepiece. When moving the eye from side to side, the two aperture images will be displaced in relation to each other. Marginal parts of the aperture diaphragm image can be obscured by the exit pupil of the objective, when the image of the aperture diaphragm is only slightly smaller than the exit pupil. This produces a decrease in brightness at the edges of the object field, apart from the above-mentioned effect of oblique illumination. The aperture diaphragm of the condenser should therefore be closed as much as is necessary to prevent the image from being vignetted, and this includes the rays which travel towards the edges of the object field.

34

1.631 Light as wave propagation. For a further discussion of the image formation in optical instruments, in particular the microscope—and for a better understanding of certain phenomena which are of increasing importance in microscopy and photomicrography, a brief explanation of the nature of light is needed. Light can be compared to a uniform sine wave form motion, when at a sufficiently long distance from the light source and over a relatively short length of observation. In order to define a wave, we need to know the wavelength λ, i.e. the distance between a point on one wave and the corresponding point on the next wave, as well as the amplitude a, i.e. the amount of movement of a point on the wave in a lateral direction.

The maximum amplitude is indicated by a_0. Wavelength λ of the light wave measured when travelling through air is characteristic for the colour of the light, and is expressed in fractions of a millimetre, usually in nm or mμ (1 nanometre or millimicron $=$ $= 10^{-9}$ metre). The value of amplitude a_0 is of practical use in as much as the radiant energy (light intensity) is proportional to the square of a_0.

The wave shown in Fig. 24 (right-hand side) can be considered both as a momentary phase of a vibration in travel (wave motion) and as a distance-time diagram of a single vibrating point. The wavelength then corresponds to the duration of a vibration, while the various amplitudes show the position of the vibrating point at any moment of the vibration. The vector on the left-hand side of Fig. 24 gives a better idea of the vibrations and vibration relationships at a certain spot, since the vibration is reduced to a central motion. The length of the vector is here value a_0. A rotation of the vector around the centre—occurring with uniform speed—indicates the operation of a vibration. The various central angles φ (the phase angles) indicate the vibration phase attained at any given moment. φ can be expressed in degrees of angle, and in radian measure as a fraction or multiple of π. The distance $a_0 \cdot \sin \varphi = a$ gives the amplitude of a point belonging to a phase.

If two or more wave trains move through a single field, the impulses emanating from the various waves will react on the individual points of the field simultaneously. Depending on their direction and size, they can increase, decrease or completely cancel out each other. This process can be expressed in a diagram by adding the amplitudes of the waves point for point according to their size and direction. Although

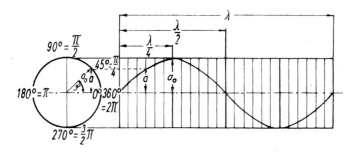

Fig. 24. A light wave shown diagrammatically.

this has a certain demonstration value, in practice this method is uneconomical. The vector diagram here is of much more value, particularly for the presentation of vibration processes at a given point of the wave field. Fig. 25 shows three cases of superimposition of two waves of equal length and amplitude at parallel vibration planes (interference). The amplitude size of the resulting wave is defined by the spatial

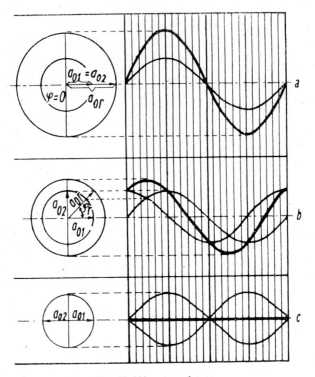

Fig. 25. Wave interference.

distance (path difference) or time distance (phase difference) of two points on the individual waves in the same vibrating position.

Two waves are superimposed in Fig. 25a, where the path difference equals 0 or a multiple of λ (phase difference 0 or a whole multiple of π). In the vector diagram, the a_0 lengths are added together. A wave of the same position, but with double the amplitude of the contributing waves, results in the wave diagram.

In Fig. 25b, the two original waves have a path difference of $\frac{1}{4}\lambda$ (phase difference of $\frac{1}{2}\pi$). The result is a wave which has a path difference of $\frac{1}{8}\lambda$ (phase difference $\frac{1}{4}\pi$) as compared with the two original waves. The vector diagram clearly shows the size and position of the resulting amplitude. In Fig. 25c, two identical waves have been assumed, with a path difference of half a wavelength (phase difference π). In this case the two waves cancel out each other.

Fig. 26 shows the diagram of a wave field in which the waves emanating from two identically vibrating (coherent) centres Q_1 and Q_2, expand and interfere with each other. The fully drawn circles represent wave crests and the dotted circles represent wave troughs. Waves with a path difference of 0 or a whole multiple of λ meet along the fully drawn hyperbolae including the axes of symmetry. The vibrating conditions

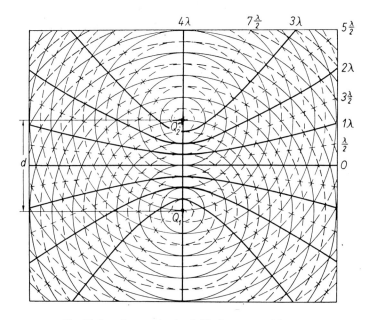

Fig. 26. Interference in the field of two wave centres.

as sketched in Fig. 25a prevail here. On the other hand, along the dotted hyperbolae, the path difference amounts to $\frac{1}{2}\lambda$ or an odd multiple thereof. A condition as represented in Fig. 25c prevails here, i.e. the field remains unperturbed.

The essential condition for interference is coherence (the same frequency and always being in phase with each other). This is easily demonstrated with models with water waves, when the surface undulates at different points, yet in the same rhythm. These waves then have identical lengths and the plane of undulation lies naturally perpendicular to the surface. The light emitted by modern light sources consists of individual, vibrating atoms, and acts as wave centres. The light emitted by them will only vibrate in the same plane during a very short time. For this reason, light waves can only cause interference when they originate from the same spot on the light source and travel along different paths, which are practically of the same length, to the field of interference. It is therefore necessary that the light emitted in a certain direction is split up into a number of part-wave trains. The images of the light source which are possibly produced, or other wave centres excited to vibration by illumination in the same rhythm, are called 'coherent light sources', and the wave trains emitted by them: 'coherent wave trains'.

Among the possibilities of producing or generating coherent part-wave trains, diffraction is the most important factor of image formation in the microscope.

1.632 Diffraction of light. According to *Huyghens'* wave theory, any point reached by an expanding wave can be regarded as a secondary centre of disturbance. When a wave expands unimpeded these secondary waves build up the front of the expanding wave. But when the wave meets an obstacle, parts of the field of wave motion which lie outside the real direction of propagation, will be caught. The secondary wave fronts now carry the motion into the space shadowed by the obstacle (Fig. 27). The intensity of the motion will decrease by increasing inclination referring to the original direction of incidence of the waves. For clearness, in Fig. 27, the reflection of the waves at the obstacle, and the propagation of the waves from the edge to the left, are not shown.

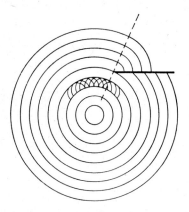

Fig. 27. Huyghens' principle. The waves reflected by the obstacle are not drawn.

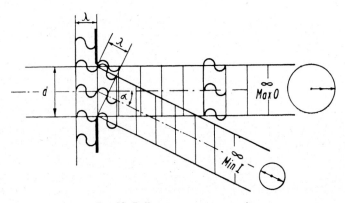

Fig. 28. Diffraction at a narrow slit.

38

In Fig. 28, a wave train coming from a great distance (and therefore not spherical but plane in contrast with Fig. 27) meets an obstacle in the shape of a narrow slit with an aperture d. A wave train of a width d will pass through the slit. The surface particles of the slit are excited to vibrate at the same time and in the same phase, and the thus formed secondary wave fronts have the direction of the incident wave as the direction of propagation, and as wave normal. The wavelets of the wave train would intensify themselves into an infinitely distant image, as illustrated in Fig. 25 a. As already mentioned, the secondary wave fronts also carry the energy into the space shadowed by the slit. The result is a multitude of propagation directions which form a larger or smaller angle α with the original direction. The direction of the wave front and of the wave normal remain, however, perpendicular to the slit. Within the inclined wave train, there is thus a deviation from the value of the angle of inclination α between the direction of propagation and the wave normal. This causes path differences. The angle of inclination α in Fig. 28 has been chosen so that a path difference of one wavelength is created between the two secondary wave fronts situated at the slit edges. In this case there exists a path difference of $\frac{1}{2}\lambda$ between the two edge waves and the centre of the slit. In the same way, each secondary wave front in one half of the slit has a corresponding partner in the other half with the same path difference. In this direction therefore, the individual waves will cancel out each other by interference, so that darkness prevails (see Fig. 25 c).

When the angle of inclination is increased, the brightness will increase and decrease periodically, resulting in brightness maxima and minima. The maximum lying in the direction of incidence is called 'maximum 0' or 'principal maximum'. It is flanked on both sides by 'minima I'. When considering diffraction at a narrow slit, the angle distance between the principal maximum and minimum I is given by:

$$\sin \alpha = \frac{\lambda}{d} \tag{17}$$

where d indicates the slit width. This is clearly illustrated in Fig. 28.

If, instead of a diffraction slit, a circular diaphragm aperture is used, a rotation-symmetrical diffraction figure will result, which is called 'Airy disc'. *Airy*'s integral is the factor 1.22, by which the dimensions of the diffraction pattern produced by a slit must be multiplied in order to obtain the dimensions of the pattern obtained with a circular aperture.

Equation (17) now becomes:

$$\sin \alpha = 1 \cdot 22 \frac{\lambda}{2 h} \tag{17a}$$

where $2h$ is the diameter of the diaphragm aperture.

When an objective is placed after the diaphragm, which unites the wave trains emerging from the diaphragm, the diffraction figure image will be formed in the focal plane of the objective, i.e. at the position of, and instead of, the geometrical-optical images of the infinitely distant light source (Fig. 29).

The image will consist of a bright centre, the maximum 0, which is surrounded by a system of concentric circles: the higher order maxima (Fig. 31).

As the diffraction figure loses its intensity very quickly, the distance between minimum I and maximum 0 is termed the radius of the Airy disc (ϱ'). We find:

$$\varrho' = \frac{a' \cdot \lambda}{2h} \cdot 1 \cdot 22 \tag{18}$$

where a' is the width of the image, here equalling F_{obj}. Fig. 30 shows the light distribution in an ideal *Airy* disc. This light distribution can only be obtained with an aplanatically ideally corrected objective, which must meet exactly Abbe's sine condition:

$$\frac{\sin \sigma}{\sin \sigma'} = k \tag{19}$$

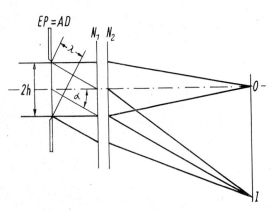

Fig. 29. Image formation of a diffraction phenomenon through an objective. The diffracting diaphragm becomes the aperture diaphragm of the objective.

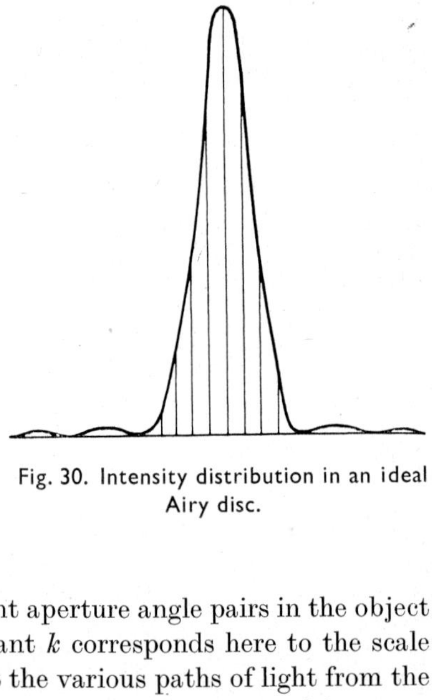

Fig. 30. Intensity distribution in an ideal Airy disc.

The sine condition states that the sines of different aperture angle pairs in the object space and image space must be constant. Constant k corresponds here to the scale of reproduction of the objective. This means that the various paths of light from the object point to the image point, should be of the same length, expressed in wavelengths. It is however impossible to meet this requirement fully, with the result that the factual light distribution will slightly deviate from the ideal image. The maximum 0 becomes slightly flatter, in favour of the subsidiary maxima.

1.633 Image formation of luminous specimens—Resolution of fine detail. Let us assume that the specimen is a luminous object, consisting of luminous points lying on a single plane. As we have seen above, even the best optical equipment is incapable of imaging individual points as such, but always as a diffraction figure, the maximum 0 of which has its area defined by equation (18). When sufficiently magnified, it can

40

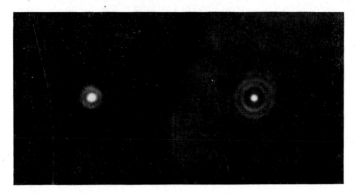

Fig. 31. Airy discs of an ultramicroscopic object, *a* focal focusing (negative plane in the geometrical-optical image plane), *b* extra-focal focusing (negative plane slightly outside the geometrical-optical image plane).

be seen as a disc instead of a point. In general, only maximum 0 is considered as an image of the object point. As its intensity decreases towards the edges (Fig. 30), the Airy disc will have unsharp borders, as seen in Fig. 31.

When two adjacent points have to be rendered as definitely separate entities, their principal maxima must be so far apart that they just touch each other. The distance between their centres is then $d' = 2\varrho'$. Substituting in equation (18), we find:

$$d' = 1{\cdot}22\,\frac{a'}{h}\cdot\lambda$$

In aplanatic systems, however, $a'/h = 1/\sin\sigma'$ because h/a', which under ideal conditions is a tangent relationship, becomes a sine relationship with aplanatic systems. The factual distance between the two object points then becomes:

$$d = 1{\cdot}22\,\frac{\lambda}{\sin\sigma'\cdot M}\qquad\left(\text{for } d = \frac{d'}{M}\right)$$

According to the sine condition, the denominator, however, is $\sin\sigma$. When a medium with a refractive index n is inserted between object and objective, the denominator will be $n\cdot\sin\sigma$, i.e. the already mentioned numerical aperture. This then makes the distance of two points which can be 'resolved' with certainty in the image:

$$d = 1{\cdot}22\,\frac{\lambda}{N.A.}\approx\frac{\lambda}{N.A.}\tag{20}$$

1.634 Image formation of illuminated specimens—Resolution of fine detail with central, oblique and dark-ground illumination. In the case of luminous objects, as encountered in microscopy when working with fluorescent illumination, each point appears as a separate light source. Illuminated objects, however, receive their light from a single light source, common to all points. This makes all the individual points 'secondary radiators', vibrating in phase with each other. The light emitted by these points is subject to interference. Interference phenomena can be observed in an image of the light source when the latter is sufficiently small. An example is halation, which can be observed when looking at a street lamp through a window covered with condensation. Another example is the uniformly arranged spectra which are seen by looking through a tissue, such as muslin stretched on a frame, or a taut umbrella. It is obvious that the appearance of the diffraction phenomena will depend on the structure of the interposed object, and that a well-ordered sequence of spectra can only be expected from a uniform structure. Such structures have the advantage that in the beginning, the relationships with microscopic image formation can most easily be understood.

The simplest structure of this type is the diffraction grating, which consists of a large number of two different types of strips in uniform, parallel arrangement. Two adjoining strips of different optical quality are termed a grating element, the width of which is termed 'grating constant d'. We distinguish between two types (Fig. 32). With amplitude or absorption gratings, the two strips of the grating element absorb the light to different degrees. In an extreme case, it consists of opaque rules and transparent slits. In the case of phase or retarding gratings, the strips of the grating elements have different refraction indici (by identical thickness) or different thickness (by identical refraction index), so that light which passes perpendicularly through the grating acquires a different phase in both parts. In the extreme case mentioned, absorption does not occur.

Let us take the case of an amplitude grating where the slit width is small compared with the wavelength of the light (Fig. 33). The grating is illuminated in a direction normal to the grating, by a narrowly limited light source, situated at a great distance. All slits will then become coherently vibrating secondary radiators. This builds up a wave front from the secondary wave fronts emanating from these secondary radiators, which travels in a direction normal to the grating. This front is represented in Fig. 33 by the tangents to the secondary wave fronts, which have no path difference with respect to each other. A maximum 0 will be formed in this direction, indicated by the arrows. A second front will be formed at each side of this principal direction of propagation, by the front of secondary wave fronts from adjacent grating elements. These have a path difference of one wavelength. One of these is represented in the diagram by a corresponding tangent, while the relative direction of propagation is indicated by dotted arrows. They lead to maximum 1 (first-order diffraction image).

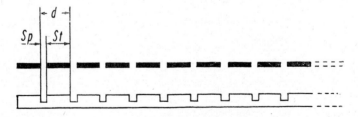

Fig. 32. Diagram of an amplitude grating (above) and a phase grating (below).

42

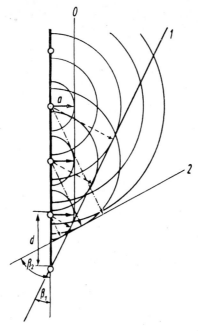

Fig. 33. Wavefronts behind an amplitude grating with very narrow slits.

One of the second-order diffraction images is indicated by the dot-line-dot arrows, tangents to the secondary wave fronts with a path difference $= 2\lambda$.

The angle distance of maxima 0 and I is defined by:

$$\sin\beta_1 = \frac{\lambda}{d} \tag{21}$$

and that of the m-th maximum of maximum 0 by:

$$\sin\beta_m = \frac{m\,\lambda}{d} \tag{21a}$$

The same considerations apply with respect to the position of the maxima to each other with gratings of a finite slit width, as in the case of the theoretical grating with infinitely narrow slits. It can of course happen that individual maxima will be lacking periodically at given width relationships between slits and rules. Moreover, the maximum 0 will gain in intensity with increasing slit width.

Contrary to the diffraction phenomena occurring in a slit, where maxima and minima pass into each other continuously, the maxima obtained by a grating are sharply defined. With light of a given wavelength, these produce sharp images of the light source. The maxima, analogous to the processes in the slit, which would in this case also be created at infinity, can in the same way be imaged by an optical system in its focal plane.

When using an amplitude grating as a microscope object, and illuminating it with a co-axial narrow pencil of light, (as can be sufficiently approximated by complete closure of the aperture diaphragm of the illumination device) the grating will produce diffraction images of the diaphragm. These will be imaged by the objective in its focal plane (Fig. 34). With this arrangement, half the aperture angle σ of the objective

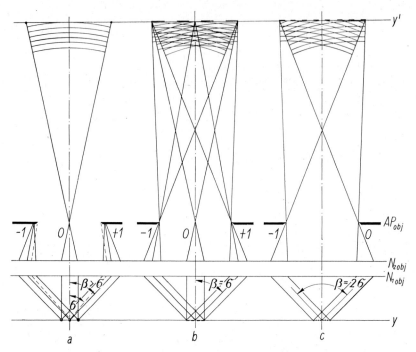

Fig. 34. The image formation of grating-like objects.

is smaller than the diffraction angle β_1, having a maximum value of 1. Only the principal maximum will pass through the objective from the diffraction figure, and will be imaged as illustrated in Fig. 34a. From here, the light will pass into the microscope tube and illuminate the focal plane of the eyepiece. In Fig. 34b, the aperture of the objective has been given such a value that the images of the two maxima 1 are imaged within their aperture ($\sigma \geqq \beta_1$). The wave fields emanating from the three maxima will now widen towards the eyepiece diaphragm and there interfere together in the object plane in the stated manner in the shape of bright and dark strips. The reciprocal distance of bright or dark strips is calculated as follows:

Before the objective (21):

$$d = \frac{\lambda}{n \cdot \sin\beta_1}$$

where $n = 1$, when the medium between grating and objective consists of air. When, after the objective, the medium is air, as is usual in microscopy, we obtain on the image side, according to the sine condition:

$$\frac{n \cdot \sin\beta_1}{\sin\beta_1'} = k = M$$

The distance between maxima I and maximum 0 on the image side is then:

$$\sin\beta_1' = \frac{n \cdot \sin\beta_1}{M}$$

It follows that the grating constant d' on the image plane is:

$$d' = \frac{\lambda}{\sin\beta_1'} = \frac{\lambda}{n \cdot \sin\beta_1} \cdot M = d\,M$$

d' has thereby as a diffraction phenomenon the same size as the geometrical-optical image of d, and replaces it. Image formation in the microscope can therefore be traced back to an interference of the coherent wavelength produced by the diffraction image.

By narrow-beam illumination with co-axial light, only those grating structures can be resolved from which the objective can receive the maximum 0 and at least the two first-order auxiliary maxima. The limit is reached when:

$$n \cdot \sin\beta_1 = n \cdot \sin\sigma = N.A._{obj}$$

The limit value for a grating constant which can be just resolved by narrow, straight illumination is therefore:

$$d_1 = \frac{\lambda}{N.A._{obj}} \tag{22a}$$

In Fig. 34 c, the illumination is effected by a narrow light beam positioned from the side in such a way that the geometrical-optical image of the light source (or the maximum 0 of the diffraction spectrum created by the grating) occurs at the edge of the aperture diaphragm of the objective. Resolution is obviously still possible when only one of the maxima 1 is produced on the opposite edge of the aperture diaphragm, since two coherent light sources are sufficient to produce interference phenomena. As the distance of the maxima can now be twice as much, the smallest value for the grating constant which can just be resolved, is decreased to:

$$d_2 = \frac{\lambda}{2\,N.A._{obj}} \tag{22b}$$

This equation, however, represents an extreme case. Depending on the obliquity of the incident light, d can take any value between d_1 and d_2. The improvement in resolving power which may be expressed as $1/d$, by decentralization, is obviously only achieved by gratings whose elements are arranged perpendicularly to the direction of decentralization. An improvement in resolving power also results as for all directions in the image field when larger illumination apertures are used (Fig. 34 b). Instead of illuminating the object with the narrow beam chosen for the experiments described above, a converging beam of finite aperture is used. This type of illumination is always used in practice. We then find:

$$d_3 = \frac{\lambda}{N.A._{obj} + N.A._{ill}} \tag{22 c}$$

Illumination methods where the principal maximum contributes to the image formation, belong to *bright field illumination* (Fig. 34 b and c).

When the illumination is so oblique that the principal maximum cannot pass through the objective, but when at least two higher-order maxima contribute to the image formation, the grating rules which so far were imaged darkly on a light background, will now appear as bright rules on a dark background. Such an illumination is termed 'dark-ground illumination'. The value of the grating constant which can just be resolved may in this case decrease to d_2.

It is characteristic of dark-ground illumination that the principal maximum is not used for image formation. Only the light diffracted from the object is utilized. It is immaterial how the principal maximum is made ineffective. This is discussed elsewhere.

The fact that the background appears bright by bright-field illumination, and dark by dark-ground illumination, indicates that the principal maximum is responsible for the illumination of the background. At the position of the images of the grating rules, the wave trains emanating from the principal maximum and the auxiliary maxima will interfere together to produce optimum reduction, while the wave trains emanating only from the higher-order maxima will exercise a mutually intensifying effect.

By illumination with finite aperture, the maxima at the edges of the objective diaphragm do not have to be fully imaged here. There must, however, be areas in various maxima, coming from the same parts of the light source (aperture diaphragm of the illumination device), since only those are coherent and capable of causing interference.

It can be proved that diffraction light produced by irregular objects fulfils the same role in image formation in the microscope as the diffraction spectra of the gratings. The same factors are therefore valid for the resolution of fine detail. As it is not possible to derive a universal equation for detail distributed at random, the resolution must always be determined with the help of the relationships prevailing at the grating.

In order to demonstrate the processes by the image formation of grating-like structures, Figs. 35—38 show photographs of a measurement grating which have been taken at different objective apertures and different systems of illumination. We refer in particular to the difference between Figs. 35 and 37. In Fig. 37, only three narrowly limited maxima from the refraction centre of the narrow grid were allowed to contribute to the image formation. It is impossible to define the width

46

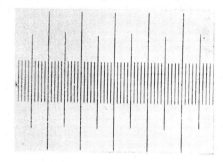

Fig. 35. Measurement graticule $M = 90:1$.
Bright field, open objective aperture.

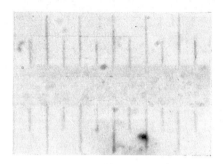

Fig. 36. Measurement graticule as in Fig. 35.
Objective so far closed that only maximum
0 contributes to image formation.

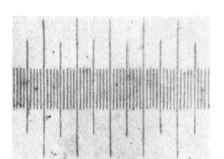

Fig. 37. Measurement graticule as in Fig. 35.
Here, maxima 0 and ± 1 were allowed to
contribute to image formation.

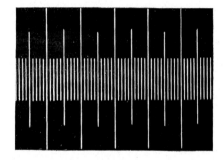

Fig. 38. Measurement graticule $M = 90:1$.
Dark ground.

of the rules and the intervening spaces, since they pass continuously from one to the other. This photograph cannot be called a correct rendering of the object. This rendering will improve the more the diffraction spectrum produced by the object structure is used for image formation. A further improvement is brought about by widening the illumination aperture to a value $N.A._{obj}$, which unfortunately decreases picture contrast. This is why mainly narrower illumination apertures have to be used.

1.635 Useful magnification. The smallest grating constant capable of resolution d_2 (22 b) must be multiplied by a certain factor, to enable the observer to see the image.

A normal eye can distinguish between two points separated from each other by at least one minute of arc. When related to the minimum distance for comfortable vision of 250 mm, this distance will be 0·0725 mm.

As some eyes are better than others, and as the observer should be able to observe object detail without strain, it is usual to take 2—4 times the above distance for the image-forming optical systems. The requirement of the angle size of the smallest object distance which can be discerned by the eye is therefore:

$$d'_{\min} = 2'\text{arc}, \quad \text{related to } l = 250 \text{ mm}: d'_{\min} = 0\cdot145 \text{ mm}$$

$$d'_{\max} = 4'\text{arc}, \quad \text{related to } l = 250 \text{ mm}: d'_{\max} = 0\cdot290 \text{ mm}$$

The magnifications required for obtaining these image sizes for the smallest grating constant which can be resolved by a microscope objective, are termed the upper and lower limits of 'useful magnification'. We shall use the symbols $Mg_{\min}$ and $Mg_{\max}$ for these concepts. By using (22 b) we find:

$$Mg_{\min} = \frac{0\cdot145 \cdot 2\,N.A.}{\lambda} = \frac{0\cdot29}{\lambda}\,N.A. \tag{23}$$

$Mg_{\max}$ will have double this value.

This equation makes it possible to calculate the limits of useful magnification for any objective and a given wavelength. For visual microscopy by white light, and for photomicrography in the central spectral range, λ is usually valued at 550 nm = $= 0\cdot00055$ mm. This gives:

$$Mg_{\min} = 527\,N.A. \approx 500\,N.A.$$

$$Mg_{\max} = 1054\,N.A. \approx 1000\,N.A.$$

1.7   Image formation in photomicrography

1.71 Classification of photomicrographic methods. Photomicrographic methods can be grouped in several ways. The most important classification in photomicrography is based on the scale of reproduction, since the latter is the characteristic value both for the type of reproduction and the necessary equipment. In Great Britain, a macro-photograph represents the object on the final print between same-size (ratio 1 : 1) and a magnification of approx. $\times 10$ to $\times 25$, when the print shows greater detail than can be observed by looking at the original from a distance of 250 mm. Assuming that photography and print production are impeccable, a same-size final print will show exactly the same amount of detail, when viewed from the conventional distance, as when the object itself is observed from the same distance. If the ratio is greater than 1 : 1, the print will show greater detail.

Reproductions with a scale more than 1 : 1 can be obtained in several ways. The one-stage imaging with the magnifying lens corresponds in macrophotography to image for-mation in the macrophotographic camera. For greater magnification than 10 or 25 times, a microscope becomes necessary, which can be designed as a camera microscope. The exact magnification where macrophotography ends and photomicro-graphy begins is difficult to determine since it depends so much on the equipment available. Macrophotography should be preferred when the aperture relationship or numerical aperture of the camera lens or microscope objective used seems to indicate this method.

1.72 The macrophotographic camera. Referring back to section 1.51 and Fig. 10, we have seen how the path of rays could be constructed by the image-forming lens in conjunction with the eye of the observer. When the eye is replaced by a photographic camera and the latter is focused on infinity, the magnifying lens will act as a photographic supplementary lens for close-ups, such as a Proxar lens. Fig. 40 shows the graphical construction of the path of rays. Only those rays which emanate from an off-axis object point, to the principal planes of supplementary lens and camera lens on the object side, are drawn, as well as the corresponding image rays.

A photomicrograph is produced when the image is considerably larger than the object and shows more detail than can be obtained from direct observation of the object from the conventional minimum distance for comfortable vision (250 mm). This gives us in the first place:

$$y' : y = M > 1 : 1$$

Two methods are available to obtain this: the achievement of the final scale of reproduction on the negative, or by subsequent enlargement of a photograph taken at a smaller scale of reproduction. With an arrangement as shown in Fig. 40, the first method to obtain the requirement will be employed when the focal length of the camera lens F_{cam} is longer than that of the supplementary lens F_{supp}, since with this arrangement:

$$M = \frac{F_{\text{cam}}}{F_{\text{supp}}} \qquad\qquad (24)$$

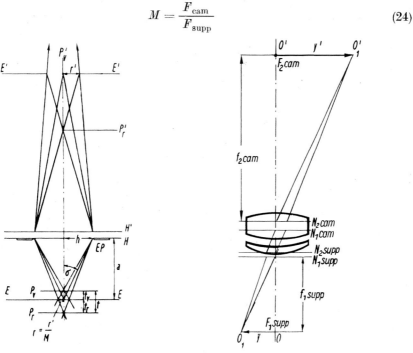

Fig. 39. Depth of field.

Fig. 40. Supplementary lens in conjunction with a camera lens of long focal length, focused at infinity.

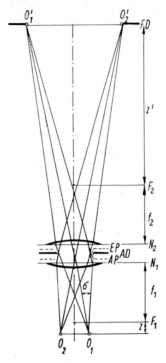

Fig. 41. Field limitation by photography with a photomicrographic lens (macrophotography).

In order to reduce the effect of optical aberrations inherent in simple supplementary lenses, their focal length is normally not shorter than 200 mm. This implies the need for camera lenses with a very long focal length, even for obtaining a magnification of only a few times. Simple supplementary lenses are therefore not ideal for making photographs at a larger scale of reproduction. Their principal application is the making of indirect small-scale enlargements, i.e. with secondary enlargement.

Special reproduction camera lenses are useful for making pictures at scales corresponding to magnifying lens enlargements (see 3.321). Fig. 41 shows the schematic arrangement. The marginal rays of the imageforming pencil of rays of the two extreme object points are shown. Within the lens, the beam of light is limited by the edge of the iris diaphragm, while in the object space limitation is ensured by the pupils (see 1.52). Apart from the brightness of the points, the value of the object-sided aperture angle 2σ decides the quantity of light which passes from these points through the lens. The function of the field diaphragm is in this case taken over by the means for size limitation within the camera, or, if this should not be present, the edge of the plate or film.

1.73 Microscope and camera. Normally, the image of the microscopic object is situated at infinity. For photography it is necessary that the image is situated at a finite distance on the focal plane of the camera. This can be achieved by arranging a camera focused on infinity in the eye position with a normally focused microscope (Fig. 42). This method of direct coupling of an optical instrument intended for visual use with a photographic camera is known in many fields of scientific photography. If possible, the two components are so arranged that the exit pupil of the first coincides with the entry pupil of the second. The scale of reproduction then depends on the magnification of the microscope and on the focal length of the camera lens, and is:

$$M_{\text{mi+cam}} = Mg_{\text{mi}} \cdot \frac{F_{\text{camera lens}}}{l} \tag{25}$$

Another method consists of removing the camera lens and producing the image direct in the focal plane of the camera, without the altering focusing of the microscope objective on the specimen, so that the intermediate image is once more formed behind the focal point of the objective. It will be necessary to extend the distance between the intermediate image and the eyepiece by an amount $(u - F_{\text{e.p.}})$, so that the pencil of rays emerging from the eyepiece converges on the various image points in the image plane (Fig. 42). When changing the distance between the two systems, one also changes the focal length of the entire microscope. According to (13a) the focal length is now:

$$F_{\text{mi}} = \frac{F_{\text{obj}} \cdot F_{\text{e.p.}}}{-(\varDelta + (u - F_{\text{e.p.}}))}$$

(5) gives then:

$$M_{\text{mi}} = \frac{k \cdot (\varDelta + (u - F_{\text{e.p.}}))}{F_{\text{obj}} \cdot F_{\text{e.p.}}} \tag{26}$$

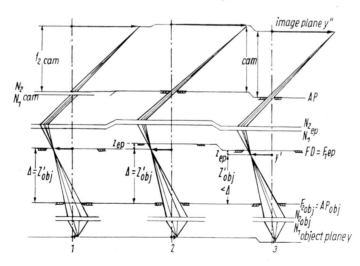

Fig. 42. Microscope and camera; positions of image and diaphragm. 1 camera with lens behind microscope. 2 focusing by pulling out the eyepiece (use as projection eyepiece). 3 focusing by keeping distance between objective and eyepiece constant Ze.p. = U — Fe.p.

where $(u - F_{\text{e.p.}})$ is linked with a camera extension k, the focal length of the eyepiece, and the optical tube length by:

$$(u - F_{\text{e p.}}) = -\frac{\varDelta}{2} + \frac{1}{2}\sqrt{\varDelta^2 + \frac{4F_{\text{e.p.}}^2}{k}} \qquad (27)$$

An approximate form of (26) is:

$$M_{\text{mi}} = Mg_{\text{mi}} \cdot \frac{k}{l} \qquad (28)$$

This equation does not take into account the change in focal length caused by extending the tube length, thus taking the eyepiece farther away from the objective. This is always permissible when working with high-power eyepieces and long camera extensions. Anyhow, it will not be difficult to ascertain deviations.

With a third method (Fig. 42), focusing is effected by changing the distance between the specimen and the objective, while the distance between the objective and the eyepiece remains unchanged. This method shortens the value of $(v - F_{\text{obj}})$ with regard to the computional value, which may impair the correction of the objective, because with shorter $(v - F_{\text{obj}})$ lengths, objectives having a higher N. A. will show under-correction. It is therefore advisable to be extremely careful when using this method in conjunction with short camera extensions (< 250 mm) and low-power

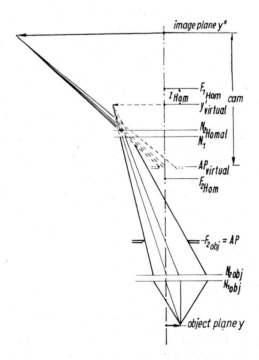

Fig. 43. Path of light when using a negative projection eyepiece.

eyepieces. The scale of reproduction can be calculated from equation (28). In order to obviate misuse of this method, the optical industry has designed special photographic eyepieces (see 3.332). They are particularly useful with camera attachments with a fixed distance from the eyepiece.

Apart from the above-mentioned systems of a positive focal length, there are also projection systems of negative focal length, e.g. the Homal projection eyepieces. Their path of rays is shown in Fig. 43. It is really the same path of rays as in Fig. 42, the only difference being that the focal lengths of the negative system have their own particular position. This forms the intermediate image produced by the objective, as well as that of the exit pupil; both are virtual images. It is therefore not possible to use a field diaphragm with a negative system. In photomicrography the field diaphragm is nearly always determined by the size of the photographic material.

As shown in Figs. 42 and 43, photomicrographic apparatus with positive projection eyepieces (including eyepieces used as a projection system), produce upright pictures. Apparatus with negative projection eyepieces yield inverted pictures, provided that no auxiliary direction-changing or image-inverting systems are being used.

1.74 Useful scale of reproduction. Photomicrographs are normally intended for visual observation. The photography must therefore be arranged in such a way that it shows full image detail when viewed from an appropriate angle. The scale of reproduction should therefore also be 'useful'. For observation from the conventional distance l, the range of the usual scale of reproduction (related to the positive) is identical with the range of useful magnification. For observation from any other distance e, the following equation is valid:

$$M_{\min e} = Mg_{\min} \cdot \frac{e}{l} : 1 \qquad M_{\max e} = Mg_{\max} \cdot \frac{e}{l} : 1 \qquad (29)$$

The useful scale of reproduction can always be obtained directly with photography on the larger sizes, such as 9×12 cm, while photographs on miniature film are usually not intended for direct observation, but are enlarged afterwards to a suitable size. If a half-plate print reveals all detail, further enlargement is pointless.

The scale of reproduction on the negative film should be chosen so that the magnification of the final positive will fall within the limits of useful magnification. In many cases, one goes below $M_{\min}$ in the negative. This does not impair the image quality since the resolving power of photographic emulsions is higher than that of the eye, so that the lower limit of the useful scale of reproduction (which depends on the photographic material) can be made lower still.

1.75 Depth of field. Specimens usually have three dimensions. It is therefore interesting to know the limitation of the depth of field, i.e. the planes in the specimen in front and below the plane which has been brought to a focus, will still be sufficiently sharp.

The depth of field is given by the reciprocal distance of the two axis points P_v and P_r, before and after the focused plane, which produce circles of confusion of a radius $2r'$ (Fig. 39). The value of r can be agreed upon, but it should be possible to see the circle of confusion as a 'point' when looking at it from a provided distance.

By geometrical-optical consideration of the relationships, a disc with a radius r' in the image plane corresponds to the circle of confusion with radius r in the focusing plane, which is cut out of the focusing plane by the aperture cones of the rays which form P_v and P_r. The basis of the cones is in the entrance pupil ($r_{EP} = h$), which, in the case of macrophotography, lies in the lens. Let us assume that the basis lies in the principal plane of the lens. The ratio that r bears to r' is the same as between object and image. Therefore: $r = r'/M$. The depth of field t consists of the measured distances in front of and beyond the focused plane, t_r and t_v.
We then have:

$$t_v = \frac{r(a - t_v)}{h} \qquad \text{and} \qquad t_r = \frac{r(a + t_r)}{h} \tag{30}$$

At high-power magnification, t_v and t_r will be so small compared with a, that they can be ignored where they appear in brackets in the above equation. Thus:

$$t_v = t_r \qquad \text{and} \qquad t_v + t_r = \frac{2r \cdot a}{h}$$

By substituting r' for r, we obtain:

$$t = \frac{2r' \cdot a}{M \cdot h} = \frac{4r'}{M} \cdot \frac{a}{2h} \tag{31}$$

where $a/2h$ indicates the ratio that the object distance bears to the diameter of the entrance pupil, the reciprocal of the 'effective relative aperture in the object space'. For practical requirements, a is replaced by the focal length:

$$t_1 = \frac{4r'}{M} \cdot \frac{F}{2h}\left(\frac{1}{M} + 1\right) = \frac{4r'}{M} \cdot m\left(\frac{1}{M} + 1\right) \tag{31a}$$

The quotient $F/2h$ has here been replaced by the 'aperture number' m. $h/F = 1/2m$ represents in aplanatic systems the sine of the largest possible aperture angle σ, and hence by photomicrographic objectives, as used in macrophotography, the engraved relative aperture or the numerical aperture corresponding with the aperture number:

$$t_1 = \frac{2r'}{M \cdot N.A.}\left(\frac{1}{M} + 1\right) \tag{31b}$$

When using these equations, we should note the way in which the apertures of the objective used are marked. With photomicrographic objectives, the engraving of the diaphragm ring with the usual aperture numbers would lead to errors, since these numbers would have only a nominal value in this case. The rings are therefore usually engraved with the aperture values expressed in millimetres. The adjusted value can in this case be used for $2h$. The last component (in brackets) occurring in the above equations must only be taken into account for smaller scales of reproduction, since its value by increasing M will increasingly approximate unity.

54

With the microscope, work is nearly always carried out at the same effective numerical aperture of the objective, unless of course a behind-the-lens iris diaphragm is used.

According to equation (15) we find that:

$$t_2 = \frac{2\,r' \cdot n}{M \cdot N.A.} \tag{32}$$

The diameter of the circle of confusion $(2\,r')$ should not be greater than 0·2 mm for images to be observed from the conventional distance. A value of 0·145 mm is regarded as effective, because, when working at the lowest limit of the useful scale of reproduction, diffraction (Airy) discs and circles of confusion will appear under the same angle of 2 minutes of arc.

As can be seen from Table 3, the depth of field expressed is very soon reduced to extremely small values. Decreasing the aperture will improve these values to some extent, but at the cost of resolution. This stopping down is done with caution in the space of the illumination light beam. Generally speaking, the purely geometrical-optically obtained equations do not give exact measurements for the depth of field, but only a basis of the dimensional relationships.

II. ILLUMINATION OF THE SPECIMEN

2.1 Methods of illumination. Fig. 44, EE represents the focusing plane in which the object is situated. Above this is the entry pupil EP of the objective. The latter is the basis of the aperture cone, which comprises the rays which image axial point O of the specimen. The specimen may appear transparent when illuminated by transmitted light, or direct reflecting in the case of incident illumination. EE is then the reflecting surface.

For the illumination of transparent specimens, the lower hemisphere of the schematically represented space is most important; for incident illumination it is the top hemisphere. Assuming that these hemispheres are mirror-like, the lower one is the mirror image of the top one. For this reason, and for the sake of clearness, the drawing has only been executed for the lower part, i.e. for work with transmitted light. All data occurring in the drawing and text can be directly transferred for conditions of incident illumination.

Rays which reach the object point from the region of the double cone between the generating lines $E\bar{E}$ and $P\bar{P}$, will illuminate this in the bright field. Illumination

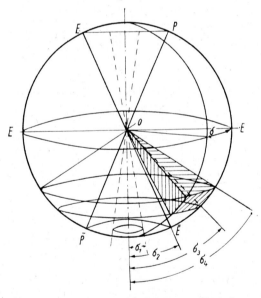

Fig. 44. Illumination and image formation in the object space. Shaded: a sector for azimuthal dark-field illumination, which has been cut out of the illumination beam with circular diameter (for universal dark-field illumination).

56

from the region outside the double cone will yield dark-ground illumination, because the light coming from these directions can only reach the objective through diffraction at the object.

The following possibilities all have practical importance:

1. Illumination cone and aperture cone of the objective are co-axial. In Fig. 44 the half aperture angle of the illumination cone (σ_1) is smaller than that of the image rays (σ_2). When looking into the microscope tube with the eye in line with the optical axis (after removing the eyepiece) the base of the illumination cone should appear concentric with the exit pupil of the objective. The illumination is then co-axial, direct, central or centred.

2. The two aperture cones are not co-axial. The base of the illumination cone will now appear eccentric to the exit pupil of the objective [same method of observation as under (1)]. The illumination is then eccentric, decentred, oblique or azimuthal.

3. The illumination cone lies outside the aperture cone of the objective. Azimuthal dark-ground illumination then occurs.

4. So far we have assumed the use of a light source having a circular diameter, such as the aperture of an iris diaphragm. With phase contrast devices, annular light sources are used, e.g. in the shape of annular aperture diaphragms. Such annular illumination also occurs when working with mirror objectives. The illumination space is then in the form of a hollow cone; the ring is always concentric with the axis of the objective. Fig. 44 shows the condition for annular dark-ground illumination, as produced by the usual dark-ground condensers. We must distinguish between a minimum (σ_3) and a maximum (σ_4) half aperture angle in annular illumination, corresponding to a minimum and a maximum numerical aperture. The difference between the two is the numerical aperture of the hollow cone. If azimuthal annular illumination is used, the illuminating ring is partly darkened.

The words 'azimuth' and 'azimuthal' have different meanings in microscopy. First, any illumination arranged asymmetrically to the objective axis is called azimuthal. This includes also irregular, non-rotation-symmetrical light intensity distribution with regard to the optical axis of the objective, as well as irregular obscuring of the light source with filters. Secondly, the concept 'illumination azimuth' indicates (in analogy with its original meaning in astronomy) the angle measured in the focusing plane between a preferred direction determined by the objective, and the principal direction of the incident light. For example, if a line-form object is so illuminated that the light passes through the object in the direction of the lines, the illumination azimuth will be 0°. An illumination normal to the lines corresponds to an illumination azimuth of 90°. Moreover, 'azimuth of illumination' may also indicate the angle subtended by the projection of the aperture angle of an illumination cone to the focusing plane. The azimuth in the case of direct illumination is therefore 360°. When a real diaphragm (azimuth diaphragm) cuts out a sector or annular section from the illumination cone with a central angle δ, the angle will have the value of the azimuth of illumination.

Azimuthal illumination causes an 'azimuthal effect' in the image. The image contours will show a more or less strong unilaterally increased contrast, depending on the direction of the contours to the incident light (illumination azimuth). The image has the appearance of a relief lit from one side. Resolution, too, varies in the different illumination directions towards the azimuth.

Apart from the above listed illumination methods, there are special techniques where a partial stopping down of the objective is used (see 6.92).

2.2   Illumination techniques

2.21 Diffuse illumination. Diffuse radiating surfaces can be either white reflectors, or ground-glass or opalglass discs illuminated from the rear. Both have the property of being able to radiate light over 180° of the object space. With appropriate shape, design and arrangement, illumination apertures of up to approximately 1.0 are possible.

With incident light, only the dark-ground radiation space is used for illumination, because the diffuse radiant surface must clear the cone EOP for the objective (see Fig. 44). This technique is used in macrophotography, when images are required which are free from shadows which are cast, and show optimum resolution. Diffuse reflectors are also used for the photography of opaque three-dimensional objects, apart from direct illumination, to brighten the shadows.

When working with transmitted light, (opal glass disc in front of the specimen) the illumination will come from the dark-ground and bright-field radiation space. Strongly absorbent object points (amplitude parts of the specimen) are imaged dark on a bright background. Phase parts, however, which in extreme cases do not absorb any light, may be suppressed completely. This technique plays only a minor part in practical photomicrography, since it is difficult to regulate the illumination aperture.

While illumination by diffuse radiant surfaces has different effects on incident and transmitted light, the following techniques have the same results for both. We shall therefore restrict our discussion to an example of the arrangement for transmitted light.

2.22 Image of light source in the objective. This technique is used for the projection of photographic transparencies, and has also proved itself in macrophotography. A lens system produces an image of the light source in the entry pupil of the photomicrographic objective (Fig. 45). The specimen is arranged closely behind the lens system. Should the beam of light at this spot not be sufficiently homogeneous (because of irregular surface intensity of the light source), a thinly frosted disc can be inserted close to the light source. The size of the illuminated object field can be limited in the case of simple arrangements only by using a real diaphragm in the immediate proximity of the focusing plane. This is, however, rarely necessary in macrophotography.

58

If possible, the light source image in the objective should not be larger than its aperture. The illumination aperture is defined by the size of this light source image.

This technique can basically also be used for photography with the microscope, when the light source can be imaged in the entry pupil of suitable objectives, i.e. imaged at infinity. A practical disadvantage will then be the difficulty of regulating the illuminated field and the illumination aperture.

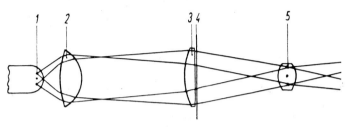

Fig. 45. Simple illumination arrangement for working with the macrophotographie camera
1 light source, *2* bull's-eye condenser, *3* spectacle glass condenser, *4* object, *5* objective.

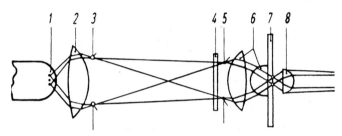

Fig. 46. Köhler's illumination. *1* light source, *2* bull's-eye condenser, *3* field diaphragm,
4 filter, *5* aperture diaphragm, *6* condenser, *7* object, *8* objective.

2.23 Köhler's illumination. This widely used type of illumination avoids the above-mentioned disadvantage. A second path of rays is here interlaced with the one producing the image of the light source in the entry pupil of the objective. This second path of rays images a field diaphragm, situated closely to the lamp in the object plane (Fig. 46).

The lens system situated in front of the light source is called a 'bull's eye condenser' and produces an image of the light source in the iris diaphragm situated underneath the substage condenser, the already mentioned aperture diaphragm. The latter, together with the image of the light source, is then imaged through the condenser, ideally at infinity, i.e. at the place of the entry pupil of the objective. The illumination device can easily be adapted to the practical conditions by means of the illumination aperture.

The condenser takes care of imaging the field diaphragm in the object plane. The field can be reduced to the required minimum size by means of this diaphragm. This avoids unnecessary illumination of parts of the object which will not be used for image formation. Another advantage is that unwanted reflections can be lessened, sometimes even completely eliminated.

A ground-glass disc for the elimination of irregular light source structure can be inserted, but this should be done in such a way that the image formation of the two diaphragms is not impaired.

2.24 Critical illumination. This technique is based on a similar use of the aperture and field diaphragms, like Köhler's illumination. The disadvantage, however, is that this type of illumination requires a light source of fairly large dimensions and uniform brightness, which is not conveniently available.

The principle of imaging the light source in the object has been basically realized in many incident light dark-ground systems, which are less sensitive to a lack of uniform brightness of the light source.

III. PHOTOMICROGRAPHIC EQUIPMENT

3.1 Light sources

3.11 Requirements of a light source. For standard photomicrography, a light source should:

1. have a minimum radiation energy (quantity);

2. be adapted or capable of being adapted to microscopic research methods with regard to spectral energy distribution (quality);

3. have a time constant of the emitted radiation with respect to quantity and quality;

4. have a minimum size and be of suitable geometric shape to be incorporated with a minimum of optical and technical difficulty;

5. be rational in consumption and produce a minimum of unwanted secondary phenomena.

Direct or diffuse sunlight gives an illumination which is not only dependent on the height of the sun in the sky and the atmospheric conditions, but will also strongly fluctuate in brightness and colour characteristics*.

The absolute brightness value of diffuse daylight is relatively low, which necessitates unbearably long exposure times, even with the simplest possible photomicrographic arrangements. The use of direct sunlight is unrational because of the fluctuations in colour temperature and the required high quality of the equipment. Moreover, the photomicrographer will be exposed to very bright light, while in the interest of increased focusing accuracy, subdued laboratory lighting should be used. These disadvantages eliminate sunlight and daylight as suitable illuminants. Electric light is therefore almost universally used for photomicrography.

3.111 Quantity of radiation. The luminosity of the light source is one of the most important factors in photomicrography. The path of rays in the instrument is immaterial, since luminosity remains constant within a given path of rays (apart from absorption and reflection). The luminosity of the light source is therefore the decisive factor for its performance in the optical instrument[2]. If an appropriate microscope lamp is used, a minimum luminosity can be calculated necessary for rational operation, which can meet given conditions, such as scale of reproduction, microscopy technique, object absorption, photographic material and exposure time. This mini-

---

* Colour temperature of sunlight before 9 a.m. and after 3 p.m. approx. 4,700— 5,800 °K
clouded sky   approx. 6,500— 7,000 °K
clear blue sky approx. 12,000—20,000 °K[1]

61

mum value, very roughly, is in the neighbourhood of 1,000 cd/cm². Light sources with a smaller luminosity either limit the range of the scale of reproduction, or require longer exposures.

3.112 Quality of radiation. The eye as well as the photographic material react to the colour temperature of the radiant energy, so that this value has to be considered when selecting a light source. The spectral energy distribution of an illuminant, which is apparent in the visible spectrum by its colour temperature, is measured in degrees Kelvin (°K) on the absolute scale of temperature, provided the illuminant emits a continuous spectrum. In the case of line radiators, the wavelengths of the principal lines are used for their characterization. The usual illuminants with continuous spectrum can be divided into two groups, according to their colour temperature: one with a colour temperature of approx. 5,200 °K, and the other with a colour temperature of approx. 3,000 °K. Illuminants with other colour temperatures should be avoided as much as possible, since they will require additional means of correction. If in doubt, the colour temperature can be measured with a commercially available colour temperature meter, and the correct filter value can be determined.[1, 4]

3.113 Constancy of radiation. Any alteration of the characteristics of the illuminant during exposure may jeopardize the result. It is well known that the characteristics of a light source depend largely on the operational voltage and current. It is therefore essential to keep these values constant. When a tungsten lamp is used, it is more important for our purpose to keep the voltage constant than the current. Fig. 47 shows the re-

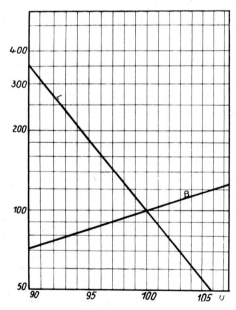

Fig. 47. Luminance B and life L of a gas-filled tungsten lamp as factor of the operational voltage U (values in per cent).

lationship between luminosity and voltage. This diagram will assist in calculating the appropriate voltage when a slight variation in luminosity is desired. We should remember that the variations in voltage also influence the life of the lamp. This pronounced dependence has therefore also been taken into account. As a rule of thumb for the variation in colour temperature it may be taken that 1 per cent variation in voltage alters the colour temperature by 10 °K. As it is most important to avoid errors caused by incorrect colour temperature, the use of a variable resistor is recommended. In the case of colour photography, for instance, the variations in colour temperature caused by voltage fluctuations should not be greater than 10 mired[67] (the colour temperature in degrees K divided into 10^6). We have, however, experienced that Agfacolor film UT 16 easily accommodates a difference of 20 mired, and UK 14 film even one of 30 mired. This can also be said of Ektachrome cutfilm, type B. Greater discrepancies will be recorded on the film as a colour cast.

3.114 Size and shape of the light source. Even today, many people are still under the impression that a point-form light source is ideal for photomicrography. This is not so. On the contrary, Köhler's illumination requires a minimum size of the illuminant, so that an optimum light performance can reach the instrument under the given geometrical and optical conditions. The size of the light source is determined by the numerical aperture of the objective and the largest useful object field, as well as by the numerical aperture of the bull's-eye condenser[3]*. The microscope lamp manufacturers have taken care of the most favourable adaption. The user should only ensure that the prescribed bulb is used in a given microscope lamp. The geometrical shape of the light source should be circular, or at least a square, while approximately identical luminosity should be present from surface element to surface element. This requirement is particularly important because as much as possible of the plane of the light source should be imaged in the entry pupil of the microscope, whilst the illumination should be of optimum homogeneity.

3.115 Economy and other factors. Economic considerations about light sources for photomicrography play an important part. The cost price here is of much more importance than current consumption. When both are taken into account, in conjunction with the useful life of the light source, i.e. the time during which the luminous flux does not decrease by more than 25 per cent, a running cost per hour can be calculated. This value should remain within acceptable limits (see Table 4). The total life of luminant (ending with its mechanical destruction) is usually considerably longer than the useful life. If the latter has been exceeded, one has to reckon with increasingly longer exposures. The values ascertained by experience for simple photomicrographic apparatus are then increased in an uncontrollable manner.

---

* The relationships are given by the simple equation: $\sin \sigma_1 \cdot R = \sin \sigma_2 \cdot r$, where σ_1 and σ_2 are the half-aperture angles of bull's-eye condenser and objective, R and r the radii of the radiant surface and the useful object field. For conventional combinations and microscope lamps, R has a minimum value of approx. $0 \cdot 6$ mm.

Finally, one of the requirements of a light source is that it does not produce unwanted secondary phenomena, in particular, strong heat and gas generation. This condition is, however, often unavoidable in contrast to the required light output, so that heat and gas generation have to be put up with.

3.12 Tungsten lamps

3.121 Lamps for general use. Lamps for general use are not particularly suitable for photomicrography. The geometrical shape of the filament gives a very low average luminosity, while a uniform illumination of objective-object field and entry pupil cannot be achieved by optical imaging of the filament. Their only application is for the illumination of a ground-glass screen placed at a distance of several centimetres from the source. This ground-glass screen then serves as a substitute light source for the microscope (see 2.21). Such simple microscope lamps with general purpose lamps are sometimes used with school microscopes, e.g. the 25 watt tubular lamp as used in the lamp shown in Fig. 49. The output is, however, insufficient for photomicrography, particularly as a considerable part of the light quantity is lost by scatter in the ground-glass screen.

Fig. 48. Tungsten lamps.

1 general purpose lamp 220 v, 25 w
2 6 volt — 15 watt projection lamp with flattened coil
3 12 volt—100 watt projection lamp.

Fig. 49. Microscope lamp.

3.122 Low-voltage filament lamps. The luminosity, so important in photomicrography, can be increased both by increasing the filament temperature, and by arranging the filament coil in a very compact form. In both cases it is advantageous to design the lamps for low-voltages[2]. The commercially available low-voltage projection lamps (Fig. 48) operate at voltages of 6, 8 or 12 volt, thereby obtaining luminosities of between 1,000 and 2,000 cd/cm² . These guarantee rational photomicrography, provided that the lamps are used in suitable microscope lamps. They ensure a good

Fig. 50. Microscope lamp with 6 v 15 w projection lamp.

Fig. 51. Microscope lamp with 12 v
100 w projection lamp.

illumination of the object field. Microscope lamps are usually fitted with an efficient condenser and iris diaphragm (Figs. 50 and 51).

The colour temperature of these lamps (2,850—3,250 °K) is correct for use with artificial light colour film. The useful life of projection lamps is approx. 100 hours, the total life 400—500 hours. Low-voltage filament lamps are reasonably economic to run, the consumption rate is low, and there are no unwanted secondary phenomena. Some low-voltage lamps have a flattened coil, and combine a considerable increase in average luminosity with the particularly interesting advantage of more uniform illumination of the object field. The improved filament design, with its closely wound coils in a very compact space, usually makes a ground-glass screen for the elimination of the filament image superfluous. This results in an additional gain in brightness.

3.13 Arc lamps

3.131 Carbon arc lamps. The carbon arc lamp has the maximum of unwanted secondary effects (enormous heat dissipation, production of nitrogen gas, constant maintenance, high running costs). On the other hand, its universality in application, and the homogeneity of the radiant surface at high luminosity, are unsurpassed by any other luminant. These points justify the use of the carbon arc lamp even to day. In photomicrography, only solid carbons are used. The light source is not the arc itself, but the glowing hollow crater at the tip of one of the electrodes (the positive electrode by D.C. operation). This crater is circular in shape and ensures a perfectly uniform illumination of the object field. The correct length of the arc is most impor-

Fig. 52. Correct position of the electrodes of a carbon arc lamp.

tant for perfect illumination. Fig. 52 shows the most favourable position when using the tip of the horizontal carbon as a light source. The average luminosity is 11,000—17,000 cd/cm² (see Table 4). Literature references disagree strongly about the colour temperature .Our own measurements with a carbon arc lamp gave 3,450 °K when operated at 10 amp A.C., and 3,850 °K when operated at 6 amp D.C. It is therefore best to take colour photographs on artificial light films. When D.C. is used, a conversion filter of type R 6 or R 3 should be inserted (see 3.612). With A.C., a very light conversion filter (R 3) may be used, but this is not indispensable.

The prescribed operational current should be strictly adhered to, since the colour temperature depends on it to a certain extent.

As the carbons burn during use, constant re-adjustment is necessary to keep the arc at constant length. The re-adjustment can be taken care of by a clockwork motor, a relay, or electric motor[5]. The two last-mentioned devices are best for photomicrography, since they ensure a high degree of constancy combined with a minimum of necessary attention. The favourable spectrum of the carbon arc lamp also makes it a suitable light source of UV, infra-red and fluorescence photomicrography.

3.132 Zirconium arc lamps (concentrated arc lamp). In concentrated arc lamps, the illuminant is a zirconium arc lamp burning inside a glass envelope. The advantages from a light-technical point of view[2] are cancelled out in photomicrography by a few drawbacks which prevent its general use. Its luminosity of 4,000 cd/cm² can almost be reached by some low-voltage lamps, whose luminous surface is nearly 30 times as large. The small luminous surface of the concentrated arc lamp requires a high optical standard. A further disadvantage is the necessity to work with D.C.

However, these lamps can be useful in infra-red photomicrography. In the short-wave infra-red region the energy of the continuous spectrum at first increases because

66

of the low colour temperature which is superimposed with a number of strong lines. The most intensive lines are at wavelengths 760, 815, 840 and 915 nm, just those regions which are most often used in infra-red photomicrography.

3.14 Gas discharge lamps. Electric discharges in certain gases or vapours of certain metals have been harnassed into light sources of very high light output. By increasing the pressure in the discharge chamber, very high luminosities can be reached. Such luminants are very suitable for photomicrography, particularly as they need less maintenance and have higher luminosity compared with the carbon arc lamp.

3.141 Xenon high-pressure lamps. Xenon high-pressure lamps can be connected to the mains. Their luminosity is between 10,000 and 23,000 cd/cm² (Table 4). They can be run at D.C. or A.C., although D.C. is to be preferred because of quieter burning and longer lamp life. The great advantage of the xenon-high-pressure lamp of type XBO (Fig. 53) lies in the fact that the colour temperature of 5,200 or 5,800 °K is independent of the supply current. A further advantage is that the lamp immediately yields its full output on ignition. After switching off, the lamp can be immediately switched on again, without having to wait for cooling off. Because of these properties, the XBO lamp can also be used for pulse operation, when an overloading for relatively long periods (maximum 3 seconds) is possible for increasing luminosity. A thrice overloading increases the average luminosity of type XBO 100 to approx. 55,000 cd/cm².

The gas discharge under high pressure produces a continuous spectrum in the visible region, which is very similar to that of mean daylight. Strong lines are present in the short-wave infra-red region at 820, 900 and 980 nm, the strongest being at approx. 900 nm. The UV region is not so well represented, so that these lamps are not suitable for either UV or fluorescence work. This is demonstrated by the fact that the ratio of blue light between XBO 100 and HBO 200 is 1 : 3, the UV radiation ratio 1 : 13.

A disadvantage of xenon high-pressure lamps is the required high pressure (maximum 20 atmospheres during operation, and 10 atmospheres when cold). This necessitates very cautious handling when changing lamps. In order not to exceed the permissible pressure, sufficient air cooling must be arranged during operation. The lamp determines the minimum dimensions of the lamp house[6, 7].

3.142 Xenon flash tubes. These are used in electronic flash equipment and have the same favourable radiation properties as the high-pressure lamps, but in a slightly modified form. The high amount of infra-red radiation of the high-pressure lamp is considerably reduced because of the absence of the glowing tungsten electrodes. This is of great advantage in microscopy and photomicrography of the physiological processes when photographing living organisms, which are not able to withstand intensive infra-red radiation. This, and the fact that the very great light energy (of the order of 10⁶ cd) only operates during very short periods (1/1000—1/5000 second), is the very real advantage of the electronic flash tubes over all other light sources of high luminosity, which would also permit the short exposure times of 1/100—1/500 second, required for the photomicrography of fast moving objects.

3.143 Mercury high-pressure lamps. Mercury high-pressure lamps of types HBO 50 to HBO 500 are suitable for some specialized fields in microscopy and photomicrography. There is a very weak continuous spectrum because of the high operational pressure of 20—70 atmospheres. The line spectrum makes it almost impossible to use these lamps for photomicrography of stained specimens by bright-field illumination, as well as by polarized light. Colour rendering and object contrast are entirely wrong, because of the lack of red light from this source. The only application in the visible spectrum in photomicrography (bright-field, dark-ground, phase-contrast and polarized light) is by using a monochromatic filter for green light of a wavelength of 546.1 nm.

Its main application photomicrographically is in the UV region (strong emission at 254 nm) and for fluorescent light work, in which case the lines in the long-wavelength UV (365 or 366 nm) and short-wavelength blue (405 nm) are exploited. For fluorescent light photomicrography, the mercury high-pressure lamp represents the most intensive and rational light source. The performance of the various types of lamps shows many differences. While the microscope lamp *L* (Fig. 54) with HBO 50 lamp (Fig. 53) is quite sufficient for subjective observation, the HBO 200 type (Fig. 53) is better suited for photomicrography. The HBO 500 type does not offer any appreciable advantage over the HBO 200 type, so that the former can be rejected because of its higher running costs.

We should note, however, that the mercury high-pressure lamp only attains its full output after 2 to 4 minutes following ignition, and can only be switched on again after complete cooling off. The high pressure makes it necessary to observe great

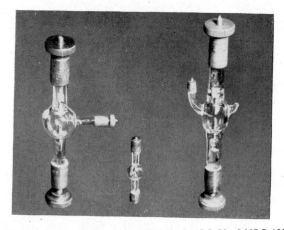

Fig. 53. Gas discharge lamps *1* HBO 200, *2* HBO 50. *3* XBO 100.

68

Fig. 54. Microscope lamp L.

caution in its handling. The lamp house must have appropriate ventilation and be splinter-proof.

3.144 Fluorescent tubes. These are mercury low-pressure lamps operated at 800 to 1,000 volts. They give a better colour rendering than the high-pressure lamps, since the inside of the tube is coated with certain fluorescent materials. The UV radiation of the mercury vapour makes these materials luminous within the visible spectrum, thereby improving both the light output and the continuous radiation. Such fluorescent tubes are not commercially available for photomicrography, but are supplied in the shape of circular tubes for macrophotography. Their luminosity of $0 \cdot 3$ to $0 \cdot 7$ cd/cm² is too small for photomicrography with the compound microscope. As the continuous spectrum is superimposed with a line spectrum, one has to reckon with incorrect colour rendering, to a greater or lesser extent, in colour photography. It will depend on the composition of the fluorescent material inside the tube whether one uses daylight or artificial light colour film. The correct choice can only be made after test exposures. Generally speaking, daylight film will give the better results. In black-and-white photomicrography, these tubes are advantageous every time larger areas are to be evenly illuminated (see Fig. 55).

3.2 Optical illumination systems

3.21 Bull's-eye condensers. Bull's-eye condensers are optical systems which concentrate the light from a source into a beam. They are frequently combined with the microscope lamp and are so arranged that the light source lies in the focal plane of the condenser, or can be adjusted to that position. By changing the distance between

Fig. 55. Actinolite in talc Incident light-dark ground $M = 2:1$.

a illuminated by 2 microscope lamps at 90°
to each other.

b illuminated by circular fluorescent tube
(glass).

light source and condenser, the emerging light beam can be convergent or divergent. The size of the condenser depends on the size of the light source. Microscope lamps are fitted with bull's-eye condensers consisting of one or more lenses, or they may be aspherical in shape, corresponding to the quality requirements of the light source imaging. Their effect can be increased by placing a concave mirror behind the light source, if the design allows this to be done.

3.22 Substage condensers. Substage condensers are optical systems which have the task of producing a uniform illuminated field of the greatest possible radiant intensity.

3.221 Bright-field condensers. We have already discussed the function of bright-field condensers (see 1.622, 2.23 and 2.24). The numerical aperture and size of the object field to be illuminated should be adapted to the optical equipment of the microscope, in particular the objective. To meet these conditions, most condensers are made so that the top combination can be removed. The complete condenser is generally used in conjunction with objectives having a numerical aperture larger than $0·40$; the lower part alone with low-power objectives having a numerical aperture between $0·40$ and $0·1$. Condensers of this type are, for example, the two-component condenser $1·2$ (Fig. 57), which is only determined by its maximum aperture, and the aplanatic condenser $1·4$ (abbreviated as Apl. $1·4$). Numerical apertures of $1·0$ and higher can, of course, only be attained if immersion oil is used between the front lens of the condenser and the underside of the microscope slide. In practice, the main difference between the two types is the quality of the field diaphragm image produced at higher apertures (see 6.211).

In macrophotography, a spectacle glass condenser is often used; this is sometimes called a 'large field condenser' (Fig. 56). The condenser should be chosen in such a way that its focal length is about 25 per cent larger than that of the lens or objective used in conjunction with it, and its free aperture corresponds to the size of the object field to be photographed. Each macrophotographic lens or objective needs its own spectacle glass condenser.

High-power photomicrography requires highly corrected substage condensers for optimum rendering of fine detail and small nuances in colour. Spherical and chromatic aberrations in simple condensers can reduce image contrast, and, in the case of colour photography, cause colour casts. These defects have been eliminated in the

Fig. 56. Spectacle lens condenser.

Fig. 57. Bright-field condenser N.A. = 1.2.

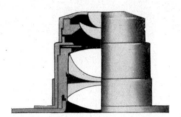

Fig. 58. Achromatic condenser.

aplanatic-achromatic condenser (Fig. 58). The superiority of this type of condenser is particularly noticeable in conjunction with objectives 60× and more. Colour photography with such objectives should only be undertaken with an aplanatic-achromatic condenser.

Pancratic condensers have a special place, as they permit a change from a small illumination field at high N.A. to a large illumination field at low N.A.

In the case of incident light microscopes, the observation objective serves also as condenser (see 6.51).

If the microscope stand has no iris diaphragm in the substage, the substage condenser must have one to enable the adjustment of the illumination aperture. The iris diaphragm is often mounted so that it can be decentred to produce oblique illumination.

3.222 Dark-ground condensers. In the dark-ground condenser, a 'patch stop' placed in the centre of the cone of light reduces the illumination to such an extent that only a hollow cone of rays emerges from the condenser, the minimum N.A. of which must be greater than that of the objective in use. The best known dark-ground condenser for high-power work is the Siedentopf cardioid condenser, a glass-mirror condenser of good chromatic and spherical correction (Fig. 59). Similar condensers are made for use with mediumpower objectives, e.g. the universal condenser shown in Fig. 60. The dark-ground condenser made by Watson is also suitable for the needs of the photographer.

In principle, a bright-field condenser can be used for dark-ground illumination by inserting a patch stop in the plane of the condenser diaphragm. The diameter of the central opaque disc must be such that its image just covers the exit pupil of the

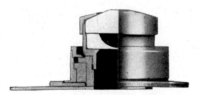

Fig. 59. Cardioid condenser.

Fig. 60. Universal condensor (for bright-field and dark-ground illumination).

objective in use. This means that each combination of objective and condenser requires its own central patch stop. These can be supplied by most microscope manufacturers. They are often laterally adjustable and rotatable for oblique illumination. The diameter can easily be ascertained by a few tests.

The pancratic condenser made by Carl Zeiss, Jena, offers a very easy conversion from bright-field to dark-ground. The focal length of this condenser is continuously variable. When combined with an annular stop as used in phase contrast work, a suitable position of the pancratic system will yield dark-ground illumination, the contrast of which does not however attain the quality of a specially designed dark-ground condenser. By varying the pancratic position, the aperture angle of the illuminating hollow cone can be varied within wide limits and be adapted to the favourable specimen rendering.

For incident light—dark-ground illumination, the condensers are frequently designed as annular mirrors, with a special one for each objective.

3.23 Auxiliary systems. Optical auxiliary systems in illumination are, for instance, lenses which affect the convergence or divergence of the beam of light between the bull's-eye condenser and substage condenser, thus shifting the light source image or the illumination field aperture. They can be used when the condensers are corrected for an infinite distant illuminant, or when the microscope lamp is arranged at a short distance from the condenser.

3.24 Beam-splitting systems. A beam-splitting system deflects part of the light beam from its normal course, when the design of the optical system of a microscope makes this desirable. Prisms, mirrors, or semi-aluminized mirrors are most frequently used. The latter kind is incorporated in microscopes for incident light, and the illumination light beam is reflected by the coating of this semi-transparent mirror. The observer views through the mirror without noticing it. The transmission of such a mirror, which may also be silvered or gilded, is approximately 50 per cent.

3.3   The microscope

3.31 Stand, stage, substage. The stand is the main part of a microscope, bearing the tube with eyepiece and nosepiece, as well as the stage and substage illumination system (Fig. 61). The coarse adjustment of the tube is actuated by rack-and-pinion movement, and should be solid enough to prevent the tube from collapsing under the weight of an attachment camera. The same applies to the fine adjustment, although modern instruments have an automatic braking device. Coarse adjustment varies with the different types of stands. We must therefore refer to the instruction booklets for the exact manipulation.

When the stand is used for macrophotography with a photomicrographic objective, the tube should be sufficiently wide to prevent unwanted vignetting. With larger

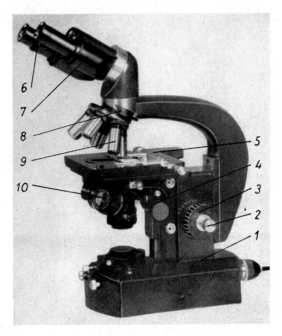

Fig. 61. Modern microscope, *1* stand, *2* fine adjustment, *3* coarse adjustment, *4* illumination device, *5* mechanical stage, *6* eyepiece, *7* tube, *8* revolving nosepiece, *9* objective, *10* substage condenser.

stands, the tube is usually fitted with an objective changing device, known as a nosepiece, which houses as many as four or five objectives. This allows the rapid changing from one objective to another, by rotating or sliding, without re-adjustment of the draw-tube.

For low-power work, the stage may be a simple manually operated, solid round or square. A mechanical stage, however, is to be preferred for medium and high-power microscopy (Fig. 61) which, in some instances, allows the object to be shifted in all directions. Graduations with vernier reading facilitate the refinding of a certain place on the specimen. A sliding mechanical stage (Fig. 62) is extremely convenient; the top plate of the stage can be slid over the bottom plate by means of an oil film. A centring stage is often required, e.g. for work by polarized light. Such centring stages are frequently provided with a graduated circular scale and a vernier (Fig. 63). Shifting the specimen can be done manually or, better, by means of an attachable mechanical stage.

The substage device houses the condenser, and with simple microscopes it consists of a sliding sleeve fixed to the stage. It is much better when the condenser carrier can be moved up and down along the optical axis. The latter is an absolute requirement for Köhler's illumination. Some substages are fitted with a centring device for the condenser sliding sleeve in order to ensure exact centring of the condenser in the optical axis of the microscope. This centring is essential when work-

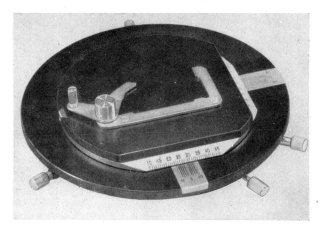

Fig. 62. Sliding stage with vernier.

Fig. 63. Centring stage with graduation and vernier.

ing with achromatic condensers of high aperture, as well as with dark-groundcondensers (unless the latter have their own centring device). Some microscope stands have a rotating and laterally adjustable iris diaphragm attached to the substage, independent of the condenser. This iris diaphragm functions as an aperture diaphragm. In this case, condensers without integral iris diaphragm are of course used.

The illumination system of modern microscopes is often integrated in the foot of the instrument. It consists of a low-voltage lamp with a bull's-eye condenser, and a field diaphragm which has outside adjusting. A mirror directs the light beam into the illumination device. This arrangement has the advantage that the illumination system is always centred with the optical axis of the microscope.

3.321 Lenses for macrophotography. Photographic lenses can be used for macro-photography when it is possible to invert their normal position in the camera (front component towards the photographic material). This is in accordance with the normal use of a photographic lens, where the object distance is always longer than the image distance. Lenses with a large relative aperture are best for photomicrography as they are usually provided with an iris diaphragm, so that the necessary depth of field can be obtained when photographing specimens which have a certain depth.

Fig. 64. Lens for macrophotography.

Special lenses for macrophotography are made by Carl Zeiss, Jena, and others Zeiss lenses go by the name of "M-lenses" (Fig. 64), available in focal lengths from 10 to 120 mm. They are anastigmatic photographic lenses, specially computed for the rendering of small objects at distances between once and twice the focal length. These lenses provide optimum overall sharpness, even at full aperture, so that stopping down to improve sharpness is not necessary for flat objects. This is, of course, in contrast with the usual anastigmatic lenses used in general photography. For the rendering of flat objects, M-lenses are therefore supplied without an iris diaphragm. The choice of focal length is determined by the size of the object field to be photographed, the desired scale of reproduction, and the available camera extension (see 6.12, 6.3, 6.6).

3.322 Objectives for photomicrography. There are dry and immersion objectives. With immersion objectives, the space between the cover glass or the surface of the specimen and the front lens of the objective is covered with an immersion medium, e.g. cedarwood oil or a suitable synthetic oil, glycerin or water. Since the immersion medium has been included in the correction of the objective, an immersion objective can only be used with the prescribed medium. Such objectives have usually a higher N.A. than a dry system of the same scale of reproduction. These objectives are sufficiently impervious to variations in the thickness of the cover glass. But the correct cover glass thickness for which a dry objective of a N.A. $> 0 \cdot 6$ is corrected, should be maintained within very narrow limits ($\pm 0 \cdot 02$ mm). Otherwise spherical over- or under-correction will occur which will make the image flat. Some objectives which are particularly sensitive to cover glass thickness deviations have the possibility of correcting this deviation within certain limits.

Objectives are also characterized in conjunction with the tube length for which they are corrected. For microscopy by transmitted light, the objectives are calculated for a certain tube length, e.g. 160 mm,* while objectives for incident light work often image the specimen at infinity and are corrected correspondingly (see 1.61). Deviations from the prescribed tube length can cause spherical errors and hence flat images. Therefore objectives and eyepieces of different manufacturers should be used together with certain precautions, i.e. with a microscope stand of a different make, since both the mechanical and the optical tube lengths are differently determined by various manufacturers.

Objectives are usually engraved on their mounts with the following data:

name and type (apart from achromats);

type of immersion [e.g. H.I. (homogeneous immersion) = oil immersion];

scale of reproduction or magnification;

numerical aperture;

mechanical tube length or image distance;

cover glass thickness.

3.3221 Achromats. Achromats (Fig. 65) are lens systems which make both a red and a blue spectral line converge to a common focal point. The residual errors for the other colours are so small that they can only be visually detected under extremely unfavourable illumination conditions. Spherically, achromats are only corrected for the medium (yellow-green) region of the spectrum, while they are over-corrected for blue rays and under-corrected for red rays. It is therefore recommended to use always a yellow-green filter when making black-and-white photomicrographs with these objectives. Otherwise it is impossible to obtain maximum contour sharpness, in particular at the edges. Achromats, and especially those of high numerical aperture, are subject to a certain amount of curvature of field.

Fig. 65. Achromatic objective.   Fig. 66. Apochromatic objective.   Fig. 67. Plano-objective.

---

* This tube length is the 'mechanical tube length', namely the distance between the objective screw-in surface and the eyepiece rest surface. It is not to be confused with the 'optical tube length' mentioned on p. 29.

3.3222 Apochromats. Chromatic aberration has been corrected for three colours of the spectrum with these objectives (Fig. 66). The residual errors occurring in the intervening wavelengths can be ignored, so that an apochromatic objective is corrected for all colours of the visible spectrum. It is therefore particularly suitable for colour photography. As apochromats show chromatic difference of magnification, they must be used in conjunction with corresponding eyepieces, which correct this aberration (compensating eyepieces). Apochromatic objectives also show slight curvature of field.

3.3223 Plano-objectives. The plano-objective (Fig. 67) is the most suitable objective for photomicrography because it has been possible to correct curvature of field, so that these objectives can produce a flat rendering of a flat object over a larger field than was previously possible. Plano-objectives can be made with achromatic or apochromatic correction. In both cases, the chromatic difference of magnification is not corrected in the objective, thus it can only be used in conjunction with compensating eyepieces. In order to receive the greatest possible benefit from the larger flat field, the use of plano-objectives in conjunction with plano-eyepieces with large object field specially computed for them, is recommended. Compensating eyepieces of older design may also be used, but they are not able to encompass the entire useful projected field.

3.3224 Mirror objectives. Simple reflecting mirror objectives (Fig. 68) only consist of a convex and a concave surface-coated mirrors. More complex mirror objectives combine the mirrors with a lens system. The advantage of this type of objective is that very little chromatic aberration is present, and that it has a relatively long unimpeded object distance. With purely catoptric systems, the positions of the principal points and focal points remain constant throughout the entire useful range of electromagnetic radiation. These system are therefore particularly important for infra-red and UV photomicrography. The lack of chromatic magnification difference prevents their use with eyepieces or projection eyepieces with compensating effect.

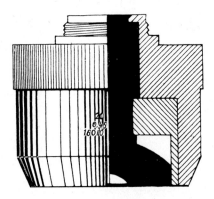

Fig. 68. Simple mirror objective. Fig. 69. Quartz objective (monochromatic objective).

Owing to the long object distance, mirror objectives have an added usefulness involving work where a greater distance from the object is desirable for thermal reasons (heating microscopy).

3.3225 Quartz objectives. These objectives are used for UV photomicrography (Fig. 69) and are specially corrected for a single wavelength, either 257 or 275 nm. This means that they cannot be used for any other wavelength. Quartz is necessary because it transmits UV radiation, while glass strongly absorbs radiations of these short wavelengths. These objectives are also called monochromats, and are differently engraved from other objectives. Instead of the scale of reproduction, the focal length is given as well as the wavelength for which the objective is corrected.

3.33 Projection systems. Projection systems are interchangeable systems which prepare the intermediate image formed in the microscope for its final use. Their task is either to magnify the detail obtained in the intermediate image according to the desired purpose, or to form an image of the intermediate image at a suitable magnified scale and at an appropriate distance.

3.331 Eyepieces. Eyepieces serve mainly for visual observation and projection of the intermediate image on the focal plane of a camera or projection screen. Furthermore they have the task to image the exit pupil of the entire microscope system at a place which can conveniently be observed by the eye pupil of the observer. Auxiliary aids, such as pointers, graticules with cross-wires or scales, are imaged simultaneously with the object and can be positioned at the place of the intermediate image where a field diaphragm is arranged in the eyepiece. There are special adjustable eyepieces available, which allow the focusing of the auxiliary aids at various distances.

Eyepieces can be divided into universal eyepieces and special eyepieces according to their function. The first variety only change the magnification or scale of reproduction of the intermediate image produced by the objective, without any optical corrections. They are mainly intended for use with low and medium power achromats, as well as mirror objectives.

Special eyepieces have specific optical properties, and are adapted for a special purpose. Special eyepieces are extremely valuable in photomicrography because they are compensating, plano and quartz eyepieces.

Compensating eyepieces are mainly intended for use with apochromats, but can also be used to advantage with high-power achromats having a numerical aperture exceeding 0·65. They compensate the chromatic magnification difference peculiar to these systems. This correction is obtained by giving the eyepieces an inverted value for chromatic difference of magnification. In the ideal case:

$$Mg\,\lambda_{\mathrm{obj}} \cdot Mg\,\lambda_{\mathrm{e.p.}} = \mathrm{constant}$$

and the projected image has hence the same magnification at all wavelengths of white light. It will be obvious that the image of the eyepiece diaphragm will show colour fringes, because this image is formed solely by the eyepiece, or part of the eyepiece. These fringes are only then objectionable when the photomicrograph of the image is limited by the eyepiece diaphragm; they can be easily eliminated with a correction filter.

Plano-eyepieces have been designed to operate with plano-objectives. The eyepiece has a compensating effect, as well as projecting a larger object field. This is because of the superior correction of curvature of field of plano-objectives.

Quartz eyepieces are specially designed for working with monochromatic objectives in UV. The term 'quartz eyepiece' originated because of its application, but it is felt that they should be called quartz projection eyepieces, since direct visual observation is not possible with UV radiation. Quartz eyepieces are computed for a projection distance of approximately 30 cm.

The designation of eyepieces varies from manufacturer to manufacturer, but always includes a few characteristic values. Universal eyepieces are designated by their magnification, and recently by a field number. The field number is the diameter in millimetres of the object field of the eyepiece. It is easy to calculate the diameter of the object field of the objective when this field number is known. At the correct tube length the diameter $2y$ of the object field of the objective is found by dividing the field number $2y'$ by the scale of reproduction M_{obj} (or $M_{obj + tube}$):

$$2y = 2y'/M \text{ mm} \tag{33}$$

Special eyepieces have additional letters for identification when the full description is not engraved. For example, compensating eyepieces are indicated by "K" and plano-eyepieces by "PK".

In addition to this, eyepieces are often designated according to their design. Two basic designs have become important for photomicrography. The Huyghensian eyepiece (Fig. 70) in its original form is mainly used as a lowpower universal eyepiece. In this type of eyepiece, the field diaphragm is situated between two plano-convex lenses, the curved sides on the side of the objective. The first lens is known as

Fig. 70.
Huyghensian eyepiece.

Fig. 71.
Compensating eyepiece.

Fig. 72.
Orthoscopic eyepiece.

Fig. 73.
Quartz eyepiece.

80

the field lens and has the task of converging the principal rays emerging from the exit pupil of the objective, so that the exit pupil of the entire system can be produced at a suitable distance from the second (eye) lens. At the same time, the intermediate image is slightly reduced in size and displaced towards the objective in the field diaphragm. The eye lens serves as magnifier for the observation of the intermediate image or projection on the final image plane. The low-power compensating eyepiece (Fig. 71) has been derived from the simple Huygensian eyepiece. The eye lens and frequently the field lens consists of cemented doublets.

With Ramsden eyepieces (including the designs derived from this type), the field diaphragm is placed in front of the lens system. The field lens lies immediately behind the diaphragm. The field and eye lenses serve the same purpose as the Huygens type. Fig. 73 represents a quartz eyepiece which comes closest to the original design.

Orthoscopic eyepieces according to Kellner (Fig. 72) have eye lenses consisting of two components. They have superior colour correction and a larger field, which is almost free from distortion. Their use as a measuring eyepiece is therefore indicated. Progress in the size of the useful field is made by orthoscopic eyepieces designed by Abbe. The lens system consists of a three-component cemented front member, followed by a simple eye lens. These lenses are positioned closely together and in this case we cannot speak of separate field and eye lenses with their specific functions. This design is especially useful in medium-power ($\times 8$) and high-power ($\times 16$) PK eyepieces.

If greater camera extensions are available, eyepieces can be used in photomicrography as projection systems without alteration. At camera extensions of less than 250 mm, care is needed, as a deterioration in correction can be expected (see 1.73). This is made particularly noticeable by the combination of a low-power eyepiece with objectives of high numerical aperture. The error can be avoided by using a longer tube, either a draw-tube or an extension tube of a calibrated set. Because of its weaker construction, the draw-tube is not recommended for use with a camera attachment. The tube extensions for a 125 mm camera extension are:

for eyepieces in the old Zeiss magnifications range:

$5\times$	$7\times$	$10\times$	$12{\cdot}5\times$	$15\times$
20	10	5	3	2 mm

for eyepieces in the new Zeiss magnifications range:

$6{\cdot}3\times$	$8\times$	$10\times$	$12{\cdot}5\times$	$16\times$
13	9	6	4	2 mm

The need for the above-mentioned tube extension does not apply to photographic eyepieces, since these can be adapted to various camera extensions by adjustment of the optical part. The positions of the intermediate image formed by the objective, and of the aperture plane are thereby maintained. By focusing the photographic eye-piece on infinity, such systems are also suitable for visual use. One can work with adjustable eyepieces in a similar way, but the correct arrangement of the optical parts must be found experimentally, since there is no scale for the camera extensions.

3.332 Projection eyepieces. Projection eyepieces are specially designed for projection and for photomicrography, where the tube extension $z_{e.p.}$ (equation 27) has been accounted for. Sometimes they are designed for a single finite projection distance, as in the case of the Zeiss MF projection eyepieces computed to function with the Zeiss photomicrographic apparatus MF. The projection distance here is 160 mm, i.e. the final image is formed 160 mm beyond the intermediate image produced by the objective, not counting image-displacing systems in the camera. The scale of reproduction engraved on the normal projection eyepiece is only valid at this single projection distance. The size of the field diaphragm, at a distance Δ behind the objective, is adapted to the image field of the camera. Projection eyepieces are available with and without compensation. They should not be used for subjective observation.

3.333 Homal projection eyepieces. This type of projection eyepiece is recommended for use with apochromatic objectives (Fig. 74). The Homal eyepieces compensate for the chromatic difference of magnification of apochromatic objectives, and are similar to compensating eyepieces. They also correct any curvature of field in the same way. The Homal eyepieces have been given a curvature of field reciprocal to that of the apochromatic objectives. Since these objectives may have curvature of field in various degrees, a range of Homal eyepieces have been computed, each to be used with a specific objective (Table 5). Homal eyepieces cannot be used for visual observation because they are negative systems and hence the exit pupil is not accessible (see Fig. 43).

A special adapter is needed to affix a Homal eyepiece to the microscope tube. It is not necessary to keep to a given camera extension. All that is necessary is to adapt the Homal II to the camera extension. A scale on the Homal's mount gives the correct adjustment (Table 6).

Fig. 74. Homal projection eyepiece.

The prime function of the camera is to carry the photographic material at a definite distance from the microscope. In the second place (but not always necessary) the camera must make it possible to focus the image. Even less important is the presence of a shutter with variable shutter speeds. One can also limit the exposure time by other methods. All that is required is to shield the photographic material in the camera against unwanted exposure. In principle, there is no necessity for the camera to have a lens, as its place is taken by the macrophotographic lens or by the microscope. It is, however, possible to use cameras with fixed, non-interchangeable lens for photomicrography.

The camera may be of the rigid type, or have variable extension. Both have their advantages and disadvantages, which are, however, a matter of purely subjective judgment. The fact that the camera with variable extension permits a free choice of the scale of reproduction at the picture taking stage is not necessarily an advantage, as the same result can be obtained by subsequent enlargement of the negative made in a rigid-form camera. On the other hand, many people think, as we do, that a rigid-form camera has the advantage of keeping the multitude of possible scales of reproduction within efficient and rational limits (see 6.12).

Among the rigid-form cameras used in photomicrography are the miniature cameras 24×24 mm and 24×36 mm, single-lens reflex cameras $2\frac{1}{4} \times 2\frac{1}{4}$ in. and the photomicrographic attachment cameras of sizes from 24×36 mm to 9×12 cm. All these cameras are intended for a projection distance of about 125 mm, measured from the exit pupil of the microscope. Rigid-form cameras for greater projection distances are only found in 'camera microscopes'. The prototype of the camera with variable extension is the bellows camera 9×12 cm and 13×18 cm. The choice of format is governed by economy, with the exception of specialized methods. From this point of view it is advantageous when the same camera can also be used for normal photographic work or close-ups outside. The normal miniature or medium-sized camera, the top models of which can meet any photographic demands from landscape photography to photomicrography, favours this point. Such a camera can, however, become very uneconomical when it is only rarely used for photomicrography. In this case, the undeniably larger expenditure in photographic material (20 or 36 exposures on a roll of miniature film) and in processing solutions, would indicate the use of a plate camera. Whether $6 \cdot 5 \times 9$, 9×12 or 13×18 cm is chosen is a matter of personal preference. The 9×12 cm camera is perhaps the most suitable for counting and measurement work on the positive print, since a contact print is large enough for this purpose, and can be comfortably seen from the conventional distance. In the region of macrophotography, or in cases when the image field of modern planoobjectives should be used to the greatest extent and without subsequent enlargement, the 13×18 cm or 4×5 in. sizes can be most useful.

In many cases of specialized photomicrography, the camera choice will be determined by the illumination possibilities, since the size of the image plane and hence the type of camera in the film plane of which the image plane is situated, has a marked influence on the exposure time, apart from light source and film speed. A simple comparison between a miniature and a 9×12 cm camera shows this clearly. Usually, the miniature frame is enlarged 3·2 times, yielding a print of $7·7 \times 11·5$ cm, which corresponds in image content (because of the unfavourable proportion of the sides only for the length dimension) to a contact print from a 9×12 cm negative which has a useful area of $8·4 \times 11·4$ cm. Because of the approximately ten times smaller area of the miniature negative, the illumination intensity in the image plane of a 9×12 cm camera is ten times larger. This means that the miniature film needs only one tenth of the exposure time of the 9×12 cm plate (the arrangement remaining otherwise identical). This is of particular importance in cases when long exposures are required by the nature of the process (fluorescence, dark-ground and polarized light photomicrography).

The $2\frac{1}{4} \times 2\frac{1}{4}$ in. size is situated between the two formats mentioned, and requires three times the exposure time of a minature film.

We can summarize camera choice as follows:

> For series work and photography under unfavourable light conditions, the miniature or medium-size camera is preferred; for single photomicrographs the plate camera.

We should reiterate that with correct processing, no difference in picture quality will occur between a photomicrograph taken with a miniature camera, and one taken with a plate camera, so that these considerations do not enter the argument at all.

3.5   Exposure devices*

3.51 Focal plane shutter. The focal plane shutter has the advantage that each point of the negative area receives the same exposure, if, of course, the source is producing an evenly illuminated object. This can be noted in pictures made for measuring the absorption or reflection capacity of the specimen. Focal plane shutters are found on miniature cameras and $2\frac{1}{4} \times 2\frac{1}{4}$ in. cameras. The shutter speeds range from perhaps 1/1000 to 1—12 seconds, as well as "B" and "T". The variation in shutter speeds is usually in steps but can be continuous.

An ideal focal plane shutter for photomicrography should have an infinitely variable range of shutter speeds from, say, 1/250 second (the shorter speeds are not used in photomicrography) to 2 seconds (longer times can be timed easily and accurately with a stop watch). It is an advantage if the speeds are infinitely variable because it is possible that the fine gradation in the exposure time, needed for the sometimes very small exposure latitude of colour material, cannot be achieved by varying the diaphragm. The shutter should have a "T" (time) setting, as working

---

* The possibility of regulating exposure by covering and uncovering the illumination by means of an opaque object has here been ignored.

with the "B" (brief time) setting and a cable release with time lock is unsafe. There should be a smooth action of the focal plane shutter. It has been shown that the shutter can have a noticeable influence on the picture quality of the photomicrograph[8]. Because of the lateral movement of the slit which starts on one side of the camera, the latter is excited to vibrate, which can lead to camera shake and unsharpness at shutter speeds between 1/1000 and 1/5 second, and scales of reproduction $> 100 : 1$ (Fig. 75). This unsharpness cannot be prevented either by vibration cushioning arrangement of the photomicrographic equipment, or by using a camera with a built-in delayed action device. Very short exposure times should be avoided when possible in the case of inanimate specimens, and the illumination intensity in the object plane decreased by means of a neutral density filter. The focal plane shutter is not suitable for the photography of live, animated objects. There are focal plane shutter cameras where the shutter action is damped to such an extent in the camera itself, that in a mechanically separated set-up of microscope and camera, no unsharpness through camera shake will occur.

Finally, the focal plane shutter should have a cable release socket. Generally speaking, any exposure in photomicrography should always be effected by means of a cable release, since pressing a release button causes a strong mechanical pressure on the camera, which can upset critical focusing.

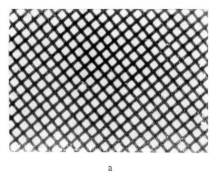

a

b

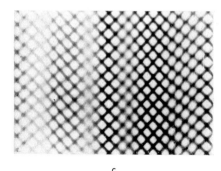

c

Fig. 75. Camera movement unsharpness in photographs with focal plane shutter cameras.
Cross-grid (grid constant 5 micron) $M = 500:1$.

a shutter speed 2 seconds.
b shutter speed $1/_{50}$ second.
c shutter speed $1/_{1000}$ second.

3.52 Lens shutters. It is based on the nature of the lens shutter that the central parts of the image field receive a stronger exposure than those at the periphery. As in photomicrography it is rarely possible to place the shutter in the plane of the entrance pupil of the microscope unequal exposure is produced at short shutter speeds. In many cases, the exposure latitude can cope with this, so that the quality of the photomicrographs does not suffer (with the exception of the special case mentioned in 3.51). Many lens shutters are infinitely variable between 1/500 and 1 second. With the use of lens shutters, the camera is much less prone to camera shake than with focal plane shutters. Unsharpness can be completely prevented when microscope and camera with lens shutter are mechanically separated (see 4.15). Self-cocking shutters, which at one time were generally used on photographic cameras, are still used on camera attachments because of their smooth action and because they do not need tensioning.

3.53 Exposure timers. Besides using a shutter, the photographic exposure can also be timed with a 'timer' which operates the light source while the cassette stays open. The advantage is that a timer and a photomicrographic assembly are entirely separate units, so that movement unsharpness cannot occur. The timers can be set to narrowly defined values, but their minimum exposure time is sometimes too long for photomicrography. The calibration cannot always be relied upon. The main disadvantage of timers is that the light source does not always follow the switching operations without inertia. However, in black-and-white photography, this may have negligible effect. As the inertia of the light source causes considerable fluctuation in the colour temperature during exposure with colour film, a colour cast may result. Noticeably incorrect colour rendering occurs with some daylight and artificial light colour films when the popular low-voltage lamps are used. Colour rendering will be correct at only an exposure of about 1 second and more. Shorter exposures may result in a red colour cast, which will be the stronger, the shorter the exposure time (Fig. 76). The timer cannot be used with tungsten lamps for exposures below 0.3 second, as the luminous flux of the lamp, and hence the colour temperature, will not have sufficient time to reach its nominal value. In the range between 0.3 and approximately 1 second, a deviation of 1/10 second from the nominal value will result in either a noticeable over- or under-exposure. On the other hand, when a shutter is used to time the exposure, the same deviation will still yield usable photomicrographs within the exposure latitude. Also in this range there is no linearity between exposure and switch-time, i.e. the timer will require longer exposure times for optimum blackening (or colour saturation) than the camera shutter, at the same optical arrangement (Table 7). The linearity is only reached at switch-times of 2 seconds and more. This drawback, and the restriction of exposure times to 1 second and longer, makes the camera shutter far more suitable for photomicrography than the timer.

Finally, we should mention that at exposure times of 1 second and longer, unsharpness through camera shake does not occur with any type of camera shutter, so that this advantage of the timer is also outweighed.

Fig. 76. Influence of the exposure timer on colour rendering—Human skin (skull). Van Gieson stain. Transmitted light—bright field; M = 63:1 original; enlarged to 250:1. *a* focal plane shutter $^3/_4$ second, *b* timer $^4/_{10}$ second.

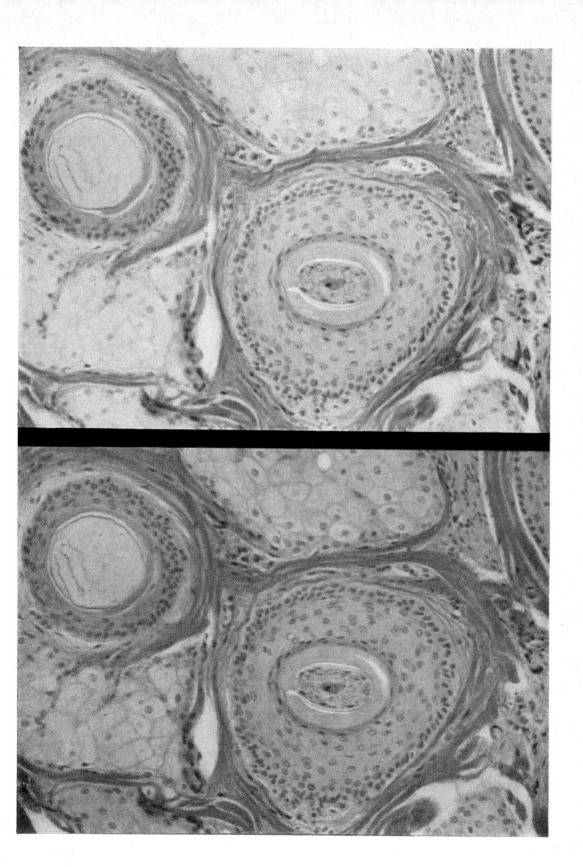

3.61 Function of filters. One can often considerably improve unsatisfactory image rendering and lack of image contrast in photomicrography (which occur particularly when working with black-and-white materials) by altering the spectral composition of the exposure light. Filters of various properties are used for this purpose. Their application is mainly determined by the function of the filter. The way in which the filtered light is produced is of lesser importance.

Any filter will cause a loss in brightness. This is of interest in photomicrography in as much as one will wish to use a filter after having made a test exposure without one. The required longer exposure can easily be calculated by means of the known 'filter factor'. This filter factor indicates the factor with which the unfiltered exposure time must be multiplied. Its value does not only depend on the light losses in the filter itself, but also on the colour temperature of the light source and the colour sensitivity of the photographic material. Table 8 gives the factors of some frequently used filters. Strictly speaking, these factors are only intended as guides, because there may be slight deviations within the manufacturing tolerance of the filters. These deviations are however so small that the calculated exposure times will fall within the exposure latitude of the photographic material.

A combination of two or more filters of different transmission will result in filters with new characteristics. An example of this is the combination of certain blue and yellow filters giving a narrow-band green filter.

The best position for all filters is in the illumination beam before striking the specimen. When placed in the image-forming beam, a high optical filter quality is essential. Here, a maximum thickness must not be exceeded, as otherwise image errors will occur. The best position is in the vicinity of the aperture diaphragm, because any uncleanness of the filter would not be apparent in the object field.

3.611 Contrast filters*. The variations in grey of the black-and-white picture of a stained or coloured specimen does not always render colour contrast as the eye sees it. Contrast filters are used to remedy this defect. Basically, any filter can be used as a contrast filter when its colour lies within the colour sensitivity of the photographic material**.

> The following guide can be used when choosing a contrast filter: specimens the colouring of which corresponds to the preferred transmitted wave-band of the filter, will be rendered light; complementary coloured specimens will be rendered dark.

---

* The classification of filters according to their function should not be adhered to too rigidly. A filter can serve several functions simultaneously. For example, a given blue filter can be adequately efficient as a colour separation filter, and can also be used as a contrast filter.

** In the visible spectrum, the filter colour corresponds to the centre of the transmitted wave-band region.

For contrast increase between differently coloured parts of the specimen, the filter should be of a colour which is complementary to one of the specimen colours. Should it be necessary to lessen the contrast, the filter should be of the same colour as the one which is too darkly rendered in the photograph. The best contrast filter can sometimes only be determined after having made several test exposures with different filters (Fig. 77). With panchromatic materials, the rendering in different shades of grey will correspond to the factual brightness values of the specimen colours, but this does not apply to the contrast with the visual impression. The use of panchromatic materials still necessitates the use of contrast filters.

3.612 Compensating filters. These are necessary in adapting the light source to the colour sensitivity of the photographic material. Yellow-green filters of a certain transmission ensure a tonevalue-correct rendering of blue-red stained specimens with panchromatic materials, e.g. eosin, azan and other stains (Fig. 78).

Daylight colour film in combination with low-voltage filament lamps needs a blue light filter to correct the colour temperature of the tungsten light. In photomicrography, this filter is mainly used for work in polarized light, but it is used for the rendering of certain colours in haematology as well. In general photography, such filters are called conversion filters.*

Neutral density filters may also be regarded as compensating filters. These filters increase the exposure time without changing the colour temperature of the light, and hence the colour rendering. They are used when the optimum exposure time falls in a region which, as in the case of focal plane shutters, would lead to camera movement unsharpness.

3.613 Colour separation filters. Colour separation filters only transmit a narrow wave-band. They are used when it is necessary to record the absorption or reflection power of a substance in relation to the wavelength of the illuminating light. When the absorption bands of the object are not too narrow, or when the reflection power does not show too much difference over larger wavelength regions, colour separation filters with relatively large transmission bands will be sufficient. These filters are available as red, deep-blue and green separation filters, e.g. Kodak 'Wratten' No. 25, 47 B and 58 respectively. Kodak 'Wratten' filter No. 18 A is opaque to the eye and transmits UV radiation (peak at 365 nm), and infra-red radiation; it absorbs all visible radiation except a very small amount of extreme red. Kodak 'Wratten' filter No. 87 transmits infra-red radiation and absorbs all visible light. Metal interference or dispersion filters mentioned in the following pages are used when monochromatic requirements are very high.

3.614 Correction filters. The colour correction of achromatic objectives can lead to a slight unsharpness of the image. A green colour separation filter will ensure optimum

---

* The mentioned filters are conversion filters of type B 14, because they change the colour temperature of the ligth source for 14 dekamired (140 mired) to the blue range of the spectrum.

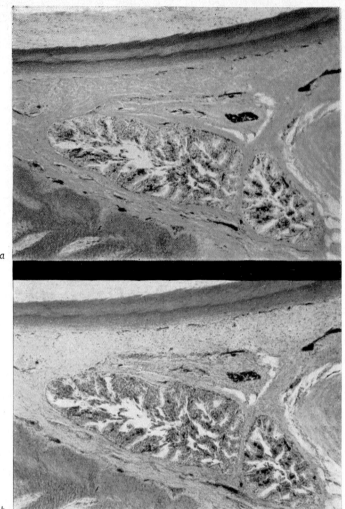

a

b

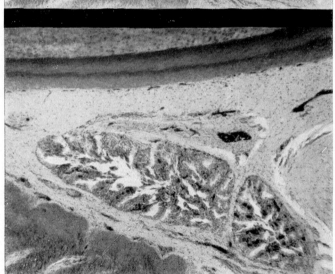

c

Fig. 77. Effect of contrast
filters.

Tongue of cock,
Goldner stain.
Transmitted light—bright
field;
$M = 25:1$ original;

enlarged to 80:1.

a Agfacolor UK 16.
b Isopan FF without filter.
c Isopan FF with blue filter.
d Isopan FF with green filter.
e Isopan FF with yellow-green
filter.
f Isopan FF with orange filter.

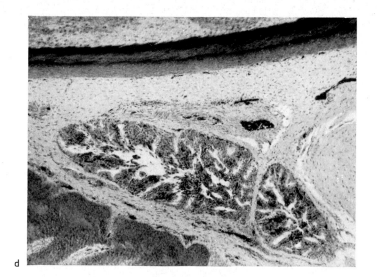

d

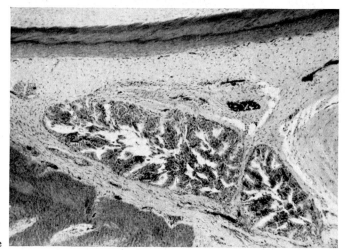

e

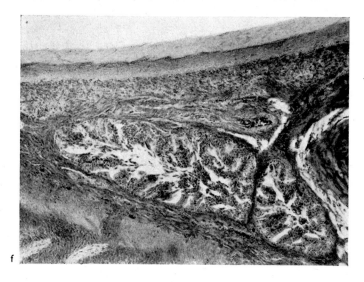

f

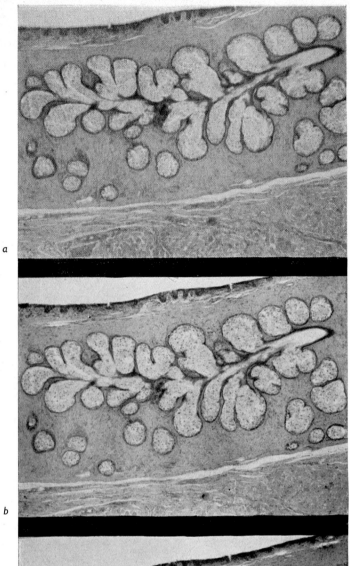

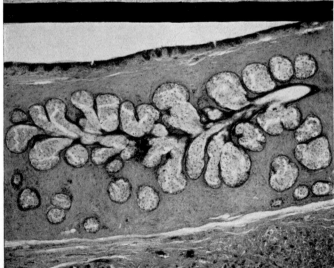

Fig. 78. Effect of compensating filter.
Skin section with tubulous gland.
Eosin stain.
Transmitted light—bright field;
$M = 20:1$ original;
enlarged to 50:1.

a Agfacolor UK 16.
b Isopan F without filter.
c Isopan F with yellow-green filter.

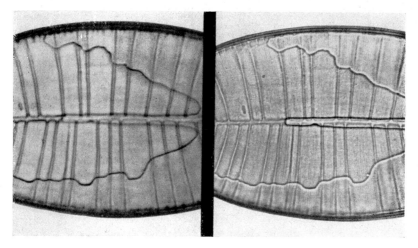

Fig. 79. Effect of correction filter on resolution. Surirella gemma diatoms; transmitted light—bright field $M = 1000:1$, left: with orange filter, right: with blue filter.

acutance. Even in the case of apochromatic objectives colour fringes may occur near the margin of the object field because of chromatic difference of magnification, if they are not sufficiently compensated by the eyepiece. These colour fringes can also be eliminated by a colour separation filter. Its colour does not need to be restricted to green, owing to the higher correction of the apochromatic objective.

We can also include under this heading filters which improve the resolving power of the microscope (e.g. the blue filter in Fig. 79).

3.62   Types of filters

3.621 Absorption filters. Absorptions filters are characterized by their property of absorbing a greater or lesser part of the incident light. This absorption is expressed by an absorption factor, the value of which depends on the wavelength of the light. In such a filter, entire spectral regions are reduced or completely suppressed, so that the emerging light is of different spectral composition.

3.6211 Gelatin filters[9]. These consist of gelatin films (coloured with certain organic or inorganic dyes) cemented between two sheets of glass. Their great advantage is that the absorption, in relation to the wavelength, can be regulated almost at will by colouring the gelatin film with dyes or combinations of dyes in varying concentrations. A drawback is that gelatin filters are not completely permanent; some may fade after long use, while all gelatin filters must be protected against excessive heat. It is this last requirement which creates difficulties in photomicrography, as it is often necessary to use light sources of high luminance and hence considerable heat dissipation, in order to achieve short exposure times.

3.6212 Glass filters. These are made from mass-coloured glass. There is a wide range of colours available, although a more restricted one than that of gelatin filters. Their use is preferred in many cases as they are stronger than gelatin filters. The transmission degree τ of a glass filter for a given wavelength λ and a thickness d can be calculated as follows:

$$\tau_\lambda = P \cdot \delta^d \tag{34}$$

where δ is the transmission degree in relation to the unit of thickness, and P a special reflection factor. The transmission degree of a combination is obtained by multiplying the separate transmission degrees. It is possible to decrease reflection losses in a filter combination by binding the individual filters together with an appropriate immersion film, e.g. the immersion oil used in microscopy. This treatment will increase the transmission degree of a combination of three filters by the quite considerable value of 0·2. Fig. 80 gives the τ_λ curves of a few glass filters[10].

3.6213 Liquid filters. These have lost much of their earlier importance, because they can be frequently replaced by the handier glass or gelatin filters. However, they have the advantage that they can be made quickly and without much difficulty by the photomicrographer himself in an emergency. They are sometimes used as heat protection filters (distilled water) but the effect of a Schott filter BG 17 of 4 mm thickness is approximately five times greater than a liquid filter of 50 mm thickness.

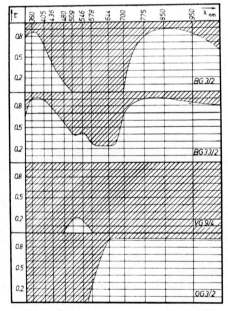

Fig. 80. Transmission factor $\tau = f(\lambda)$ of a few glass filters.

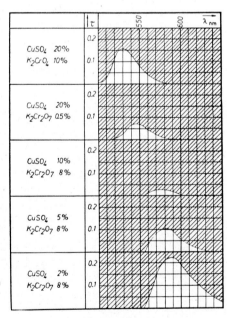

Fig. 81. Transmission factor $\tau = f(\lambda)$ of Köhler's liquid filter combinations.

Liquid filters with 1—3% copper sulphate solution are used in fluorescence photo-micrography for the absorption of red light. These filters can, however, be replaced by glass filters, e.g. Schottglas BG 12.

We should mention a very interesting combination of liquid filters. By combining two cells, one filled with copper sulphate solution, and the other with either potassium monochromate or potassium bichromate, a filter will be obtained which can filter out any desired portion of the medium visible spectrum by varying the concentrations of the individual solutions. *Köhler* recommends the following concentrations for cell thickness of 50 mm:

Copper sulphate in distilled water	20, 10, 5, 2 and 1%
Potassium bichromate in distilled water	8, 2, 0·5%
Potassium monochromate in distilled water	10, 1, 0·1%

The solutions have excellent preserving properties[11]. All the above concentrations are not required in practice. Most of them yield more or less saturated hues of green. A filter combination which has the best monochromatic properties within the 500—650 nm region, is represented by the τ_λ curves in Fig. 81. Other concentrations will only alter the degree of transmission and the band-width, while maintaining the transmission centre.

3.622 Interference filters. Interference filters are used as monochromatic colour separation filters, and are able to isolate an extremely narrow band from the spectrum of a light source. This is rarely needed for photomicrography with visible light, but is most useful in UV and infra-red photomicrography.

3.6221 Plane-parallel interference filters. These consist of a transparent, optically operative film of a certain thickness, coated on both sides with translucent films of metal. Through multiple reflection and interference of the light, separate wave-bands are filtered out, while a very steep decrease in transmission degree occurs in their immediate proximity (Fig. 82[12]). The transmission maxima which occur at given wavelengths lie at distances which can be approximated by $\lambda_1 : \lambda_2 : \lambda_3 = 1 : \frac{1}{2} : \frac{1}{3}$. The subscripts 1, 2 and 3 also represent the order of interference. To eliminate unwanted secondary phenomena, the interference filter is cemented with appropriate absorption filters.

Values λ_{max}, τ_{max} and the half-value width HW are characteristic for a monochromatic filter. λ_{max} indicates the wavelength at which the transmission degree reaches its maximum value τ_{max} at any desired position of the spectrum. With interference filters, the value of τ_{max} is between 0·30 and 0·40. Finally, the half-value width HW represents the difference between wavelengths at which the transmission degree of the filter has reached half its maximum value (approximately 10 nm). Strictly speaking, all these values are only valid for perpendicularly incident light. When possible, these filters should be inserted in a parallel beam of light, or at least

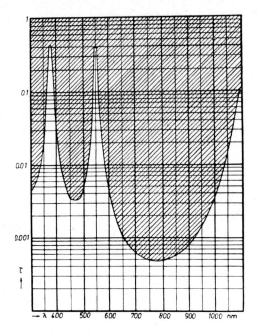

Fig. 82. Transmission factor $\tau = f(\lambda)$ of an interference filter.

a beam of light with a convergence of less than 15°. This is not difficult to achieve in photomicrography, when the interference filter is inserted in the proximity of the field diaphragm.

Interference filters are available for wavelength regions of $330-1{,}100$ nm and have extreme filter factors which are determined not only by the high selectivity and the relatively low τ value of the filters, but which also strongly depend on the colour sensitivity of the photographic material used, according to the position of $\lambda_{\max}$. Table 8b gives the filter factors for a few important line filters. The values are self-explanatory.

3.6222 Graduated filters. Recently it has become possible to make interference filters with graduated thickness, having a linear relationship between $\lambda_{\max}$ and the position of the light passage in the graduated filter. With the Zeiss graduated filter, for instance, the maximum transmission wavelength shifts almost linearly from 400 to 750 nm over a distance of 70 mm. The maximum transmission (0.40) and half-value width (10 nm) remain almost constant throughout the entire range. These favourable properties make the graduated interference filter of particular interest as a monochromator for photomicrography, especially as the type of monochromator used in spectroscopy can hardly be employed because of its lack of light intensity. An interference monochromatic filter can replace an entire filter set, since the image contrast can be quickly changed by simple sliding of the filter and when adapted to the prevailing

96

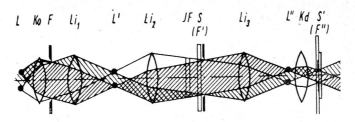

Fig. 83. Path of rays in a monochromator with graduated interference filter
$L(L'L'')$ Light source and its images Ko Bull's-eye condenser.
$F(F', F'')$ Field diaphragm and its images, Li_1, Li_2, Li_3 Auxiliary lenses for intermediate images of L
and F, IF graduated interference filter, $S(S')$ slit and its image, Kd condenser.

conditions by the best possible method. When a limited amount of monochromation is sufficient, it is only necessary to place the graduated filter closely below the moderately opened aperture diaphragm of the microscope. For higher requirements, it will be necessary to isolate individual wave-bands by means of a slit (Fig. 83). The slit and field diaphragm are made optically conjugate, so that Köhler's illumination principle is valid. At a slit width of 1 mm, the half-value width of the monochromator is approximately 15 nm. If the objective object field is larger than the slit image on the objective plane, illumination can be obtained by inserting a rotating prism behind the slit.

3.623 Dispersion filters. Dispersion filters are rarely used, although they have been used in photomicrography as monochromatic filters. A dispersion filter consists of a dense filling of powdered optical glass in a cell of a given thickness. The spaces between the powder particles are filled in with a liquid. Glass and liquid are so chosen that their dispersion curves intersect at the wavelength which is to be isolated. At this wavelength, the refraction indici of both glass and liquid will be the same, so that the filter acts as a homogeneous optical medium and transmits light. Light of any other wavelength will be reflected or bent (according to the differences between the refraction indici) and disappear from the light beam. The transmission curves of dispersion filters are similar to those of interference filters, but in the case of dispersion filters it is possible to isolate a narrow band, together with a very considerable increase in transmission degree. *Korolew* and *Klementjewa*[13] give the half-value widths of 3—4 nm at a maximum transmission degree of 0·8—0·9 for dispersion filters. Because dispersion curves are dependent on temperature, it is possible to vary the filter centre λ_{max} by varying the temperature. Dispersion filters are inserted in a parallel beam of light and can be used for infra-red[14], visible and UV radiation.

3.624 Gaseous filters. Gaseous filters are used in conjunction with liquid and glass filters for the isolation of short-wavelength UV radiation from the spectrum of mercury lamps. The monochromation of the combination is not very high. The band filtered out comprises the lines between 250 and 300 nm, and yet it is just this very region which is so important for absorption research in biology. *Meyer-Ahrendt*[15] has given a filter set for the photomicrographic presentation (within certain limits) of the distribution of purin derivatives and amino acids in tissues. The set consists of an appropriate filter, a solution of 51 gm nickel sulphate heptahydrate in 100 cc water in a 30 mm thick quartz cell and a gaseous filter of chloric gas which is pumped into a 50 mm thick quartz cell, and to which 2 or 3 drops of liquid bromine are added before sealing.

IV. SETTING UP PHOTOMICROGRAPHIC APPARATUS

4.1 Photomicrography with the microscope

The simplest photomicrographic apparatus consists of the separate groups: microscope, camera and microscope lamp. Since a certain amount of adjustment is demanded for impeccable work, it is recommended to arrange the separate groups so that no coarse disarranging can occur during work. This can be achieved by having a T-shaped distance part, which is often supplied with the microscope lamp (Fig. 50). The positioning of the camera differs from case to case.

4.11 Cameras with fixed lens. A copying stand is required to connect the camera and microscope. The stand has an adjustable carrier to which the camera is attached by means of its tripod bush. The camera lens axis should coincide with the optical axis of the microscope (Fig. 84). The camera lens should be set to infinity (∞), and should operate at such a height above the microscope eyepiece that the diameter of the beam of imageforming rays emerging from the microscope is smaller than the operative diaphragm aperture of the camera lens. Otherwise, vignetting will cause

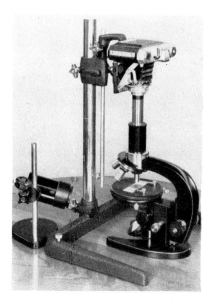

Fig. 84. Camera with fixed lens in conjunction with microscope.

parts of the image to be cut off (Fig. 85). The best position is achieved when the exit pupil of the microscope coincides with the entrance pupil of the camera lens, and the lens diaphragm is opened fully.

Such a set-up will yield a scale of reproduction dependent on the focal length of the camera. The scale of reproduction will be considerably smaller than subjective enlargement (see 6.1221). This reduction factor, which is usually $\frac{1}{5}$ in the case of minature cameras, makes it impossible to use the full negative size. The image is rather limited by the eyepiece diaphragm (Fig. 86). This is advantageous when it is required to record photographically the entire subjectively visible eyepiece object field.

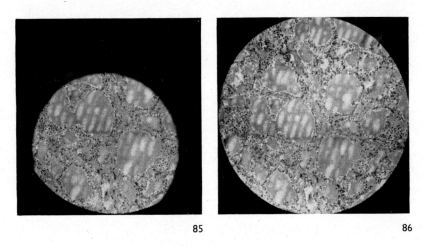

85   86

Fig. 85. Vignetted photomicrograph caused by wrong adjustment of camera to microscope with the arrangement of Fig. 84; Thyroid gland; eosin stain, transmitted light—bright field, Rolleicord IIa with Triotar f/3.5, 75 mm, M = 72:1; reduced to 63:1.

Fig. 86. As Fig. 85, but with correct adjustment of camera.

4.12 Cameras with micro adapter. With cameras with interchangeable lenses, the lens is replaced by a micro adapter, often supplied by the camera manufacturer. This adapter can be used with any eyepiece (Fig. 87). The length of the adapter is given by the distance between the image and exit pupil of the microscope, and hence the scale of reproduction. Some of the micro adapters commercially available are unfortunately rather short. The distance mentioned above should not be smaller than 125 mm, or pronounced spherical aberration will result. As most micro adapters have extension tubes (as used for close-ups), the total length can be adjusted by means of the insertion of further extension tubes. When a distance of 147 mm is observed between the upper edge of the eyepiece and the focal plane of the camera, projection eyepieces as well as high-power eyepieces can be used without impairing picture quality. The micro adapter is placed on the microscope tube in such a way that the latter's upper

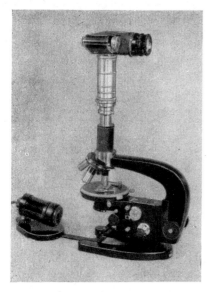

Fig. 87. Single-lens reflex camera, micro adapter and microscope.

Fig. 88. Single-lens reflex camera, nearfocusing bellows device and microscope.

edge is hidden by the lower part of the adapter. In this arrangement, the camera should have focusing means, or a ground-glass or clear-glass focusing screen of the same overall length should be available. Modern single-lens reflex cameras allow the use of different types of focusing screens. Instead of a plain ground glass (matt field lens), focusing screens with a clear-glass circle or ring, or even clear-glass screens with cross-wires, can be used. This choice of focusing aids facilitates focusing in photomicrography, but does not eliminate the general disadvantage that with this arrangement the entire field recorded by the camera cannot be surveyed. The so-called wedge rangefinder cannot be used, because the aperture of the image forming pencils of rays is too small, so that in each case only half of the rangefinder wedge can be fully illuminated.

4.13 Miniature camera with special stands. For some cameras, special close-up stands with a bellows near-focusing device are supplied. They are excellent for use with the microscope. The near-focusing device is coupled with the microscope by means of a lightproof sleeve, without touching each other (Fig. 88). By changing the bellows extension, it is possible to select the most favourable image part. For practical work it is recommended to mark the column with two lines, one for a camera extension of 125 mm (camera factor 0·5) and one for an extension of 250 mm (camera factor 1).

101

4.14 Attachment camera. The attachment camera is perhaps the most rational device for the rapid production of photomicrographs. As the name implies, the camera is placed on top of the microscope and securely attached to it. Its main characteristic is that the camera can work blindly, as a special focusing system has been built into the attachment camera. Fig. 89 gives a schematic representation of the functioning

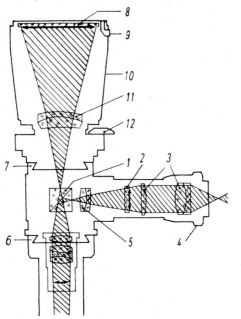

Fig. 89. Path of rays in attachment camera.

1 beam splitter.
2 focusing screen.
3 focusing optical system.
4 focusing ring.
5 auxiliary lens.
6 microscope tube with projection eyepiece.
7 base frame.
8 photographic plate.
9 cassette holder.
10 attachment camera.
11 image plane adjustment lens.
12 shutter speed disc.

of such an apparatus. A special tube is used on the microscope, at the same time carrying the base of the attachment camera with the focusing eyepiece. A beam-splitter is incorporated in the base which deflects approximately 20% of the light emerging from the microscope into the focusing eyepiece, while 80% reaches the film plane. A total reflecting prism is sometimes used instead of a fixed beam-splitter, which can be swung in and out of the focusing position. An auxiliary lens images the object image on a graticule with both cross-lines for focusing and a format-indicating frame. In this case, the frame is for 24×36 mm and $2\frac{1}{4} \times 2\frac{1}{4}$ in. (Fig. 90). The focusing eyepiece is focused on this graticule. As graticule and film plane are optical conjugates, a sharp image will also be formed in the film plane.

The attachment camera has some remarkable features:

1. Focusing is aerial by means of a clear glass and cross hairs. This ensures a higher focusing accuracy and a higher image brightness than is possible with a ground-glass screen.

2. The object field in the focusing eyepiece is larger than the picture frame size of the camera in use. This is more advantageous because with the set-ups described

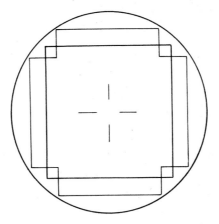

Fig. 90. Focusing screen with graticule.

in 4.12 and 4.13, the image field can be smaller than the frame size. This is due to the design of the single-lens reflex camera, where the relatively small mirror area cuts off part of the image in the direction of the shorter dimension (this design is necessary to allow the use of wide-angle lenses).

3. Attachment cameras with a beam-splitter permit the specimen to be watched continuously, even during exposure. This is very important when photographing moving objects, because one immediately knows whether another picture should be taken because of object movement.

Fig. 91. Attachment camera $2^1/_4 \times 2^1/_4$ in.

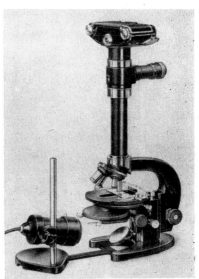

Fig. 92. Exakta Varex camera with phase contrast microscope.

4. As the image is viewed through the eyepiece, the attachment camera remains permanently attached to the microscope, which greatly contributes to the immediate readiness for photography.

Attachment cameras are usually designed so that, apart from the camera attachment made by the microscope manufacturer (Fig. 91), any miniature camera can be used with a special adapter, provided the camera lens can be removed (Fig. 92).

4.15 Stand cameras. With regard to the larger special photomicrographic units, the vertical arrangement has proved to be the most popular for working by transmitted light. The Leitz Aristophot and Zeiss Universal cameras are typical examples. They consist of a solid base plate which carries the column for the camera carrier, and the microscope lamp. The latter is usually provided with a low-voltage filament lamp, which can be exchanged for another type of light source, e.g. fluorescent light. Its axial alignment is assured by parallel guides. The microscope is placed on the base plate and its position locked after co-axial adjustment with the camera. This permanently guaranteed correct alignment of lamp, microscope and camera ensures trouble-free working. Any type of camera can be used—from the 9×12 cm bellows camera (Fig. 93) to the 24×36 mm, $2\frac{1}{4} \times 2\frac{1}{4}$ in., or $6\cdot5 \times 9$ cm attachment cameras (Fig. 94).

With the bellows camera, an infinitely variable adjustment of the extension is possible. The lower end of the bellows carrier takes the shutter and a lightproof collar. This collar is designed so that no physical contact between camera and microscope occurs, while being completely lightproof. This is important because otherwise adjustments of the camera would inevitably be transmitted to the microscope. The top of the bellows has a carrier for the focusing screen (ground glass or clear glass) and/or the cassette and plate. A mirror housing for observation of the image from

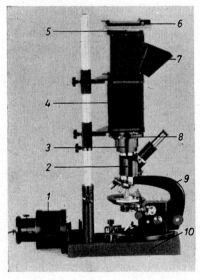

Fig. 93. Large-format camera and microscope.

1 microscope lamp.
2 interchangeable tube for photomicrography.
3 light-tight sleeve.
4 Bellows camera 9×12 cm.
5 multiple exposure attachment.
6 cassette.
7 reflex attachment.
8 shutter.
9 microscope stand.
10 base plate.

Fig. 94. Medium-format camera and microscope.

1 attachment camera.
2 reflecting eyepiece and focusing device.
3 camera support.

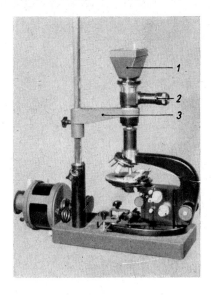

the operator's sitting position (vertical set-up) can also be attached as well as a multiple-exposure attachment for determining the correct exposure time.

After removing the bellows, a carrier arm can be attached to the column, which has a lightproof collar. A quick-changing device takes the attachment with the chosen camera, which may be a miniature camera with adapter, or a camera attachment. The mechanical separation between microscope and camera in conjunction with the robust carrier arm assists in obtaining a sharp image, even with small camera formats, with the exception of focal plane shutter cameras with harsh shutter action.

Binocular viewing is often preferred for research work requiring only infrequent photography. The camera should, however, be available for immediate use at any time. Special tubes are obtainable for this purpose which allow a rapid change from observation to photography. For instance, the Zeiss Universal Camera is available with a 'photo tube' with connections for both projection eyepiece or Homal eyepiece, and a device for visual observation (monocular or binocular). Switching from observation to photography and vice versa is effected by insertion or removal of a totally reflecting prism. Contrary to what takes place with an attachment camera, the image planes are not conjugates in this case, since the bellows camera has a variable image distance.

4.2 Camera microscopes

A camera microscope combines microscope, camera and illumination in a single unit. Its main advantage is constant readiness for use. Another feature is the permanent alignment of the light source within the microscope, and the latter's connection with the camera. This type of equipment is extremely robust in design, and guarantees extensive vibration-free work. Some types are mounted on a special vibra-

Fig. 95. Large incident-light camera microscope.

tiondamping base (Fig. 95) and allow photography at a scale of reproduction ranging from 0·5 : 1 to 1,300 : 1. Miniature cameras can also be used.

The camera microscope illustrated in Fig. 96 is specifically designed for transmitted light operation, and has an integral film chamber for the 24×36 mm format.

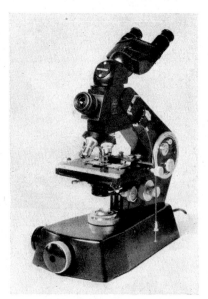

Fig. 96. Transmitted-light camera microscope in position for photography.

V. THE PHOTOMICROGRAPHIC LABORATORY

Sometimes the photomicrographic equipment is situated in rooms where other activities are carried out at the same time as photography. A rational practice of photomicrography demands the observance of a few guides which will now be discussed.

Rooms which are subject to vibration caused by street traffic and electric motors in other parts of the same building, are not suitable for photomicrography. Neither are rooms where there is a considerable number of persons moving about. Even with relatively short exposures, it will be difficult to avoid unsharpness. If no other room is available in the building, the equipment should be installed as free from vibration as possible. The vibration dampers shown in Fig. 95 are very useful. If unobtainable singly, foam rubber underlays will render the same damping surface, particularly those of 30-40 mm thickness fitted on the bench carrying the equipment. This method will yield excellent freedom from vibration, even for high-power work. It is sometimes necessary to adapt the thickness of the foam rubber pad to the weight of the equipment.

Preparatory photomicrographic work can be carried out by normal laboratory illumination. It is fortunate when the level of room illumination can be varied and adapted to the image brightness on the focusing screen. It is then possible to carry out both preparatory and photographic work at the same eye adaptation. Frequent changes in eye adaptation can cause premature strain, and hence less accuracy in focusing. The room illumination can be varied simply by opening or closing dark window curtains.

Some attention should be paid to the electric installation because the devices mentioned here can considerably facilitate photomicrographic work and are soon compensated by the gain in working hours. Apart from the usual mains supply of 230 or 240 volts A.C., a D.C. supply can be very useful when carbon arc lamps are used. It should be noted that A.C. arc lamps cannot be operated with D.C., as this would require a different type of carbons as well as a special regulating device. Gas discharge lamps burn more quietly with D.C. than with A.C., at the same time increasing the lamp life to 800 hours as opposed to 500 hours. Both carbon arc lamps and gas discharge lamps need a magnetic mains auto-transformer which gives a constant voltage with a maximum deviation of 4% for a mains deviation of approximately 15%. This deviation is permissible for photomicrography. Stronger fluctuations cause primarily changes in intensity, but can also lead to failures because of changes in colour temperature. A rectifier is needed for D.C. supply.

Generally speaking, a rheostat or variable transformer will render the same services, but requires constant control by the operator. Its advantage lies in the fact that a low tension can be applied during visual observation, and the full tension for photography. This method naturally increases the lamp life. Lamps can also be over-run, but only for black-and-white photography (because of the change in colour temperature) and at the price of a shorter life.

Heat, gas and ozone generated by gas discharge and carbon arc lamps require ample ventilation for their dissipation. The laboratory should always have air conditioning, or at least a window. Motors for air conditioning apparatus should be installed as far away as possible from the laboratory. The photomicrographic equipment should not be exposed to very strong light. Windows, or strongly reflecting walls should be neither in the viewing direction of the operator, nor in that of the focusing screen (see 8.13). The best place is parallel the wall opposite the window, with the unobscured part of the window at the back of the operator.

VI. WORKING WITH PHOTOMICROGRAPHIC APPARATUS

6.1 General considerations

6.11 Requirements of the specimen. The mounted specimen sometimes causes errors in the reproduction, which can be avoided by careful preparation of the slides. Foremost are spherical errors, caused by an unsuitable thickness of the cover glass (see 1.531). When choosing the thickness of the cover glass, it should be remembered that a thin film of mounting medium will lie between the section and the cover glass. Experience has shown that this film has a thickness of between 0·01 and 0·02 mm. Therefore the thickness of the cover glass should be that much thinner. The mounting medium must not be wedgeshaped, because this could also impair picture quality. Any striation in the cover glass or mounting medium could cause astigmatism in the image field.

Metallurgical photomicrography sets a very high standard for the evenness of sections. This is even more important as unevenness in the reflecting surface will have the same effect as a faulty mirror. Curvature of field and partial unsharpness in the image can frequently be attributed to unevenness of the specimen. If sections are uneven, the depth of field formula will be helpful, thus working out the two extreme planes, above and below the focused plane. The same errors occur with inaccurate microtome sections.

It is obvious that the mounting medium must be free from any foreign matter. When cleaning mineralogical sections with a cloth, any fragments of textile fibre clinging to the specimen would make themselves objectionably noticeable when crossed polarizers are used. The advice to dye histological specimens a little more strongly than usual for colour photography, given by many authors, cannot always be followed. Rather, one should try to obtain optimum results from the photographic material by control of the taking conditions. One piece of advice can, however, always be followed: namely the use of an embedding medium which is free from absorption, and which is as transparent as possible. It should never give a yellowish impression. On the other hand, experience has shown that canada balsam, which is frequently used in petrography, has no perceptible influence on colour rendering in the very thin cement layers used in this technique $(10-15\,\mu)$. But thicker layers would almost certainly cause a yellow cast. Detailed information on slide preparation will be found in the literature[16-20].

6.12 Selection of scale of reproduction

6.121 General considerations. The choice of a scale of reproduction is invariably a question of rendering the finest object details so that they can be clearly observed in the final print when viewed from the conventional distance. It is doubtful whether the ratio (image size to object size = 1) is the most rational method of characterizing this increased resolution of fine object detail. The concept of 'actual resolution' introduced by *Grabner*[21], in reality does not convey more than the scale of reproduction, since it only characterizes the performance of the microscope or photomicrographic apparatus with relation to the smallest object detail to be rendered. It does not give any information about the relationship to the other structures rendered in the photomicrograph with regard to their size. It is, however, useful to employ both concepts together (scale of reproduction and actual resolution). With the given value of scale of reproduction it is possible to get the ratio of dimensions in the object from the photomicrograph, while the scale needed for the rendering of fine details of a certain distance can be easily calculated from the relationship between scale of reproduction and actual resolution (see table 9).

The scale of reproduction is usually determined by making it equal to the scale required for the visual observation of the desired structure. This is only valid when contact prints will be made from the negative, generally when working with large formats. If further enlargement of the negative is envisaged (35 mm) the scale of reproduction for the negative should be correspondingly smaller (e.g. divided by 3.2 when enlarging a 24×36 mm negative on 9×12 cm paper).

With photomicrographic apparatus having an infinitely variable scale of reproduction (bellows cameras, apparatus with pancratic projection systems) the size of the part of the object to be recorded can be varied, particularly when the aesthetic aspect of a photomicrograph is important. When measuring the scales of reproduction thus obtained, values such as $263 : 1$, $718 : 1$, $1,127 : 1$, etc. will frequently be found. These values are therefore entirely determined by subjective impression. Attempts at standardization have resulted in the standard numbers R 10 in DIN standard 323. There is a fixed relationship between the individual numbers of the standard range: the logarithm of the standard number N is a whole multiple of 0.1 ($\log N = n \cdot 0.1$, with $n = 0, 1, 2$, etc.).

After conversion to the tenth power, the same numerical values will re-appear. These standard numbers are used in Table 9. The advantage of a standard range is obvious, namely, by restricting the infinite number of possible ratios to the values of this range, similar photomicrographs produced by different authors are directly comparable. A further advantage is that each multiplication or division of a standard number will yield another standard number, so that, despite initial prejudices, working with these values is easy and rational. The graduation of the range enables the selection of the most suitable scale of reproduction for any given specimen.

6.122 Calculating the scale of reproduction. We shall now give a summary of the equations derived in section 1 for the various photomicrographic combinations.

6.1221 Cameras with fixed lens. The scale of reproduction on the photographic material is obtained by multiplying the scale of reproduction, or magnification number of the objective, with the magnification of the eyepiece, and the focal length of the camera lens divided by 250 (expressed in millimetres) according to (25):

$$M_{\text{picture}} = M_{\text{objective}} \cdot Mg_{\text{e.p.}} \cdot \frac{F}{250}$$

110

6.1222 Cameras with micro adapter and camera attachments. For a camera extension $k = 125$ mm, the scale of reproduction (when eyepieces are used) equals half the subjective magnification (28):

$$M_{\text{picture}} = M_{\text{objective}} \cdot Mg_{\text{e.p.}} \cdot \tfrac{1}{2}$$

When a camera attachment is used, the scale of reproduction must be multiplied by its camera factor K:

$$M_{\text{picture}} = M_{\text{objective}} \cdot Mg_{\text{e.p.}} \cdot \tfrac{1}{2} \cdot K$$

If the length of the micro adapter is not 125 mm, the factor $\tfrac{1}{2}$ in the above equations must be replaced by $k/250$.

Projection eyepieces give by maintaining the prescribed image position:

or
$$M_{\text{picture}} = M_{\text{objective}} \cdot M_{\text{proj. e.p.}}$$

$$M_{\text{picture}} = M_{\text{objective}} \cdot M_{\text{proj. e.p.}} \cdot K$$

6.1223 Cameras with variable extension. The scale of reproduction of the photographic material is calculated by means of equation (28), by measuring the camera extension k from the exit pupil of the microscope in millimetres:

$$M_{\text{picture}} = M_{\text{objective}} \cdot Mg_{\text{e.p.}} \cdot k/250$$

In the case of a Homal eyepiece we obtain:

where
$$M_{\text{picture}} = M_{\text{objective}} \cdot M_{\text{Homal}}$$

$$M_{\text{Homal}} = \frac{\text{diameter of the image on focusing screen}}{\text{field number of Homal}}$$

The field diaphragm must be opened sufficiently wide to prevent any vignetting of the image area, nor should there be any vignetting diaphragms in the path of rays.

By one-stage image formation (macroscopy) the scale of reproduction is determined according to equation (7) by image width v, measured between the lens diaphragm and the film plane, and by the focal length of the lens F:

$$M_{\text{picture}} = \frac{v}{F} - 1$$

Using a miniature camera with interchangeable lenses and extension tubes, this equation becomes more simple:

$$M_{\text{picture}} = \frac{Z'}{F}$$

where z' is the increase in extension compared to the position of infinity of the lens.

6.123 Measuring the scale of reproduction. By calculating the scale of reproduction according to 6.122 one obtains only approximate values, since the actual values of $M_{\text{objective}}$, $M_{\text{proj. e.p.}}$ and $Mg_{\text{e.p.}}$ may deviate from the engraved rated values. Exact measurement of k is also rather difficult, as the exact position of the exit pupil of the microscope is hard to define. *The most accurate method of determining the scale of reproduction is by means of a micrometer slide.* The rulings are usually 0·01 mm apart. Measurement is carried out as follows: Firstly obtain a sharp image of the micrometer ruling on the focusing screen of the camera, with the selected combination of objective and eyepiece; then measure the length of the greatest possible number of rulings with a ruler, or, preferably, a slide gauge. When the distance of *n* rulings is measured to be *a* mm, the scale of reproduction is found when using a micrometer slide with 1 mm divided into *x* parts:

$$M = \frac{a \cdot x}{n} : 1$$

It is obviously not rational to carry out this time-consuming measurement, which also involves exchanging the specimen for the micrometer slide every time that a photograph is taken. Much better to determine the scales of reproduction for all required objective-eyepiece combinations once and for all, and keep them in the form of a table. In the case of miniature photographs with a fixed extension, the secondary

FIG. 97. EXAMPLE OF A TABLE USED FOR DETERMINING THE CAMERA EXTENSION FOR A DESIRED SCALE OF REPRODUCTION

Scale of reproduction	Objective	Homal	Division mark	Position of pointer*	
				approx. value (mm)	value measured on the instrument
100:1			(7)	510	510
125:1	Apo ×15	II	(8)	635	636
160:1			(10)	800	801
200:1	Apo ×15	VI		655	651
250:1				790	784
250:1			(8)	580	580
320:1		II	(10)	735	735
400:1			(12)	910	910
	Apo ×32				
400:1				605	603
500:1		VI		725	722
630:1		II	(10)	790	790
630:1				550	550
800:1	Apo ×60			655	654
1000:1				785	785
		VI			
1000:1				560	554
1250:1	Apo ×90			670	668
1600:1		IV		530	520

* The values given are for use of the camera without the multiple exposure attachment.

112

magnification factor can immediately be read off the table, which thus becomes an ebact standard value of the print scale of reproduction. The bellows guiding rod can xe graduated, or a correspondingly long ruler with pointer can be arranged parallel to the rod, with cameras with variable extension. It is then necessary to note in the table the values for the camera extension (measured along the graduated rod or ruler), at which the corresponding objective-eyepiece combination gives the desired scale of reproduction. This method not only saves time, but is also accurate (within the given measurement uncertainties) and is therefore to be preferred to other methods, e.g. establishment of a factor table, in which, for each combination, the scale of reproduction in relation to the camera extension is represented in the form of a curve. Fig. 97 gives an example of the above-mentioned table.

6.13   Choice of optical equipment

6.131 Colour correction of objectives. The required image quality will determine the choice between achromatic and apochromatic objectives which is not decided by the technique used (see 6.9). Because of better colour correction, apochromats, even without narrow-band filters, yield greater sharpness than achromats. Moreover, the line definition in the image centre is also better with apochromats than with achromats. This distinction is only applicable when it is necessary to render extremely fine object detail as sharply as possible. The resolving power is naturally the same for both types of objectives, provided that they have the same numerical aperture. Apochromats are therefore the choice for the highest quality, while the less expensive achromats are preferably used for routine work.

6.132 Combinations of objectives and eyepieces. The scale of reproduction required to render desired specimen detail decides the numerical aperture of the objective to be used. The N.A. should ensure that the scale of reproduction of the final print (including the secondary magnification factor) lies within the useful range (see 1.74). As there is an empirical relationship between numerical aperture and scale of reproduction[21], the objective is normally determined through the prescribed N.A. according to its scale of reproduction or magnification; thus the scale of reproduction of the projection system is also determined (taking into account the secondary magnification factor). In spite of this, the results may be ambiguous, as illustrated by the following example. For optimum rendering of structure and morphology, an object needs a scale of reproduction of 400 : 1. From the available range of achromatic objectives, achromats 20/0·40 and 40/0·65 would both be suitable. A scale of reproduction of 400 : 1 lies within the useful range for both (lower limit with achromat 40 and upper limit with achromat 20). The problem of which is the best objective in this case is therefore not solved by the N.A. alone. The following rule can be applied for making the final selection:

If the image centre should show optimum line definition, the objective must be chosen so that the total scale of reproduction lies at the lower limit of the useful range. If, however, uniform image sharpness over the entire image field is required, it is better to work near the upper limit of the useful range. The latter case is also valid for plano-objectives, because unwanted colour fringes at the margin of the field are eliminated. If at a given scale of reproduction optimum depth of field is required, it will be necessary to choose the objective with the lowest possible numerical aperture.

Once the objective has been selected in the described manner, the projection system follows on from the calculation formulae in section 6.122.

It is more advantageous to work with micrographic lenses at a scale of reproduction up to about 40 : 1. These have the advantage of superior colour and spherical correction over a combination of a low-power achromat and an eyepiece. Also, curvature of field is less than the marginal unsharpness given by a low-power apochromat in conjunction with an eyepiece.

6.14   Final focusing

6.141 Cameras with fixed lens. The camera lens is set to the infinity mark and the microscope focused so that the image formed by the objective and eyepiece is at infinity. When focusing the image observed in the eyepiece, with a relaxed eye accommodation at infinity, a sharp image will also result on the focal plane of the camera when placed in position. Practice has revealed that many people have difficulty in using a non-accommodated eye. In order to minimize this influence, a small pocket telescope which has previously been adjusted at infinity, is used for critical focusing, by placing it on top of the eyepiece. The telescope is removed before the camera is placed in position.

6.142 Ground-glass screen focusing. Single-lens reflex cameras, camera attachments and bellows cameras, all permit focusing on a ground-glass screen in the film plane. This has the advantage that the entire image field can be seen in a single glance, and that (e.g. in the presence of curvature of field) focusing can frequently be effected so that the sharpness appears to be uniformly spread over the image. A drawback to the ground-glass screen is the loss of image brightness according to the degree of frosting; also, the ground-glass grain makes focusing of very fine image detail rather difficult. Additional magnification of the ground-glass image by a focusing magnifier is of little use, since the grain of the screen is equally magnified. Focusing is then only for minimum unsharpness.

6.143 Clear-glass screen focusing. Focusing on a clear-glass screen is more efficient than ground-glass focusing, particularly at low image brightness. Here the focusing magnifier (Fig. 98) is most advantageous (Mg = $3-6\times$). This magnifier is designed so that it is possible to screw the lens system in and out for adapting it to the eye of the observer. Without any ground-glass grain, focusing of extremely fine object detail is easier and more exact. On the other hand, the focusing magnifier now also produces an image of the exit pupil of the microscope, in which position the observer should place his eye. The pupil image lies relatively far from the lens, a fact which necessitates some training at first. When the eye is not in the correct

114

Fig. 98. Focusing magnifier.

position, the observer will see either nothing at all (by eccentric pupil position) or only part of the image field encompassed by the magnifier (by co-axial position and incorrect distance).

The clear-glass screen itself only serves to determine the image plane. For this reason it has cross-wires on which the focusing magnifier must be focused. Again, it is important that this focusing should involve as little eye accomodation as possible. The best way is to focus the magnifier against a uniformly illuminated or luminous surface. The lower part of the magnifier is placed over the cross-wires and the screen is held against the bright surface. The magnifier is now moved downwards in its mount by relaxed accommodation of the eye, until the cross-wires appear sharp. This focusing position is not changed afterwards. The eye which is not used to look through the magnifier fixes on an object preferably at least 8 feet away. Short-sighted observers should wear their corrective glasses here, as well as during the later work. This will facilitate focusing of the maginifier for the infinity-accommodated eye.

When the magnifier, similar to an eyepiece, is an integral part of the instrument (attachment camera) and possibly has a graduation in diopters, focusing on the cross-wires has to be done first, before the object is focused. The small telescope mentioned in section 6.141 can here be of considerable assistance. The image field should have ample brightness and as little structure as possible. Spectacle wearers (with the exception of astigmatics) can correct their eyesight with the focusing magnifier, i.e. work without glasses. A warning, however, is necessary about indiscriminate use of the diopter scale found on the mounts of many focusing magnifiers to compensate for faulty eyesight. This graduation is only meant as a reference scale, and the engraved values only approximate the real refraction indici. On the other hand, this scale helps to ascertain that focusing has been carried out in the correct way. If, for instance, a normal-sighted observer finds that the focusing magnifier indicates a value of, say, — 5 diopters after focusing on the cross-wires, he can be certain that he did not use relaxed, accommodation-free, eyesight.

The correct position of the image in the focusing plane can be checked by means of the parallax test. When the eye behind the magnifier moves sideways or upwards slightly, the sharply focused object on the clear-glass screen should not move against the cross-wires. If it does, the magnifier focusing, and hence the after-accommodation of the eye, should be altered until the parallax between object image and cross-wires is no longer present.

We have devoted so much space to this point owing to the great importance attached to focusing, especially with a magnifier. Any error imaged will inevitably be translated by an unsharp image in the image plane—and onto the film or plate, since clear-glass screen and image plane are not optical conjugates in this case.

Systems which produce residual curvature of field require the sharpness to be evenly distributed over the entire image field. In this case, focusing is not carried out in the image centre, but at points at a distance of about one-third of the image field diameter from the edges.

6.15   Choice of photographic material

6.151 Black-and-white materials. For optimum object recording, the specimen should not be prepared as for microscopy by staining, which is not possible with live objects or with photomicrography by using polarized light, and rarely with histological specimens. On the contrary, the negative material must be adapted to the specimen and to object detail[22]. The best rule is to change the type of film or plate as little as possible. Restriction to a few types of high-contrast and low-contrast film or plate is quite feasible. If the object rendering requires optimum adaptation by means of the negative material, every desired gradation can be obtained by these two types by varying the filters and development conditions.

With regard to colour sensitivity, an orthochromatic emulsion is more useful than a panchromatic. It is exactly the lack of red-sensitivity which makes orthochromatic materials ideal for producing contrasty pictures of specimens with red-stained components. Panchromatic emulsions are mainly used in special techniques (see 6.9). Thin-film plates and films are normally used in photomicrography because of their high resolving power. This automatically excludes high-speed materials of 100 ASA and faster, as these are coated with thick emulsions. Another reason for using slow materials is their strongly reduced halation, which guarantees high acutance as well as their fineness of grain, a quality which is maintained by the use of fine-grain developers. This is particularly important for 35 mm film work. The only disadvantages of miniature film in photomicrography are granularity (caused by incorrect processing) and reduced sharpness (caused by halation). The image contenst of the miniature negative in relation to object detail rendering, will always correspond to the resolving power of the microscope based on the numerical aperture of the objective in use, and the wavelength of the light used for illumination.[23]

When miniature negatives will be subjected to high after-magnification, the most important thing to avoid is granularity. The best material for this purpose is document copying film developed to gamma 1·0. But this film does not offer any advantages

116

within the useful range of scale of reproduction, compared with other thin-film emulsions. Their inherent steep gradation is rather a disadvantage, so that document copying film cannot be recommended for universal use[24]. A film that could be recommended is a thin-film material with a speed of about 8 ASA in tungsten light (e.g. Agfa Isopan FF).

With regard to plates, the choice is wider than with miniature films, and it will depend on the work to be done most of the time which type of plate is chosen as 'universal' material. The Kodak P300 panchromatic plate (40 ASA for tungsten light) is excellent for the microphotography of coloured or stained biological, mineralogical, metallurgical and palaeontological specimens. This is thus an all-purpose fine-grain plate of high contrast, and is available either unbacked or with an antihalation backing which clears during development. It is sensitive to UV and to the visible spectrum. The Ilford R.40 plate has similar characteristics and is also available with and without antihalation backing. Plates with antihalation backing shoold be preferred. The Agfa Micro plate has high resolution and excellent freedom from halation. At extremely low brightness levels in the plate plane, reciprocity failure will occur, requiring an exposure factor of between $2 \times$ and $4 \times$, when the exposure has been calculated according to section 6.162. Other methods (with the exception of automatic exposure meters) take reciprocity failure into account automatically.

The photographic material should be fresh. Fog must not exceed $0.1-0.2$, as a higher value would have an adverse effect on print exposure time and gradation. It is a good idea to buy as much photographic material of the same emulsion number as is likely to be used within the stated period of freshness (date indicated on the packing). Working with plates and films of the same emulsion batch facilitates objective determination of the exposure time, since each change of emulsion may mean a new calibration of the exposure meter. As, on the other hand, fog should not exceed a given maximum, a well-estimated buying of material is very important in photomicrography, which only experience brings. Outdated materials must never be used; they will only lead to disappointing failures. Table 10 gives a list of black-and-white materials suitable for photomicrography.

6.152 Colour materials. Colour reversal films are mostly used in photomicrography because of their superior colour rendering. It is up to the photographer to have a good knowledge of this material, as well as an excellent mastery of the photomicrographic apparatus and taking technique, to obtain optimum results. Faulty colour rendering caused by wrong processing hardly ever occurs, as the processing laboratories work with sufficient and constant accuracy. In the case of negative colour film, the position is entirely different. Apart from the fact that the positives made from the negatives usually show considerably less colour saturation than reversal film, there are many more possibilities during printing, of 'correcting' the colour rendering. Only the photographer himself can obtain the best results, since only he knows the colours of the original. This will hardly ever be possible to arrange, but it is a good plan to take pictures on reversal colour film at the same time as taking exposures on negative colour film. It is suggested to send the reversal

117

film results to the processing laboratory which processes the negative film, to serve as a colour guide. The negative-positive process should only be used when prints on paper are absolutely essential, e.g. in the case of handwritten examination work. Blockmakers prefer transparencies for making blocks for colour printing, as obtained with reversal film.

Another question which arises when choosing between different colour films is for which type of lighting the film is balanced, daylight (approx. 5,600 °K) or artificial light (approx. 3,200 °K). The light source used will determine which type must be employed (see 3.1), although in some cases experience can provide the answer (see 6.976).

The resolving power of colour film is generally 40—50 lines per millimetre, which is acceptable.

In the chapter on black-and-white materials we mentioned the advantage of working as much as possible with films of the same emulsion number. This applies even more to colour films, because there are generally small unavoidable differences in 'balance' between films of different emulsion numbers In the case of black-and-white film, any discrepancy in speed can be rectified by adjusting the exposure meter. In colour photography, however, light-balancing filters must be used, such as the Kodak 'Wratten' filters series 81 (brownish, to lower the colour temperature) and series 82 (bluish, to raise the colour temperature).

If a test exposure shows a strong colour cast, there is little point in trying to correct this by means of a filter. The result will never be perfect. Therefore, should the difference in colour temperature exceed the values of the light-balancing filters, it is better to use a film with a different emulsion number.

In the same way we can eliminate reciprocity failure effects in the various colour layers. These faults also appear in photomicrography in emulsions which would yield impeccable results under normal conditions of general photography. It is possible to compensate this to some extent by using a light source with higher luminous intensity.

The fact that, in spite of accurate work, we sometimes get wrong colours (particularly by illumination with relatively monochromatic light or a number of pure spectrum colours) is due to the nature of the integral subtractive colour film, and cannot be avoided.

Additional special points to be considered when choosing and using colour materials are discussed in section 6.9 in conjunction with the corresponding photomicrographic method.

6.16 Determining exposure time. In photomicrography it is essential to determine the optimum exposure time, and not just any value which lies within the exposure latitude. This is necessary to obtain optimum blackening of the negative. There are several methods for determining the exposure; the one chosen will depend on the available time and on economic considerations.

6.161 Test exposure. This used to be the only way of determining correct exposure. Even today, it is often practised, mainly with single exposures. When a plate is used, it is exposed strip by strip, so that each strip receives double the exposure of the preceding one. The correct exposure is determined after the developing and fixing of the test plate. The 'real' exposure is then made with this exposure time, but great care should be taken to ensure that the taking conditions have not changed between test and real exposures. Miniature film photomicrographs can be exposed in the same way, and the best of 4 or 6 frames can be selected.

Test exposures on plates can be made in two different ways. One method is to effect the division in strips by means of the cassette slide, which should bear a division as shown in Fig. 99. This division separates the central portion of the plate into seven sections of equal width. The loaded plate holder is inserted into the camera and the slide pulled out to line 0—0. After having estimated the approximate range of exposure times needed for the specimen in hand, the first exposure is given (1/100 second, 1 second or 1 minute), indicated in strip P. The slide is then pulled out to the next line (1—1/100) and the exposure indicated in the next field is given. And so on, until all seven exposures are completed. The individual exposure times can be added; the total exposure time can be read off strips A (see Table 11). This

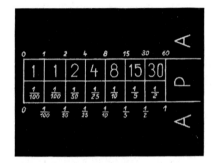

Fig. 99. Division on the cassette slide for test exposures.

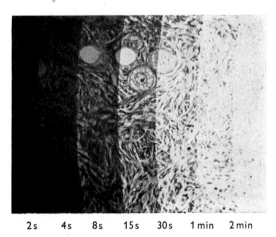

2s   4s   8s   15s   30s   1 min   2 min

Fig. 100. Test exposure; exposure in strips by moving the cassette slide, cat's ovary, eosine stain, transmitted light—bright field; $M = 200:1$.

method has two disadvantages. First, a different part of the object is reproduced on each test strip (Fig. 100). It could be that the optimum exposure time found in the fifth strip is not necessarily the best one to dissolve object detail present in the third strip. This drawback can be ignored for objects without pronounced absorption or reflection differences. Secondly, the effect of the intermittency effect could be felt. The blackening effect produced by a number of short exposures is not the same as that produced by the equivalent single exposure. In photomicrography, however, the intermittency effect can be ignored.

More advantageous is taking test exposures with the multiple exposure attachment, (available for 9×12 cm (Fig. 101) and 13×18 cm bellows cameras, as well as the device shown in Fig. 102) which is placed on top of the camera and is provided with a mask which allows a strip of 13×80 mm for a 9×12 cm camera to be recorded on the plate. The plate holder and plate can be slid over this mask, so that a series of test exposures can be made with the plate holder at different click-in positions (Fig. 103). The slide is removed from the plate holder.

Fig. 101. Multiple exposure attachment
9×12 cm with slit mask.

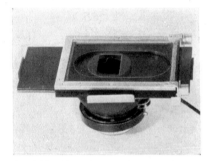

Fig. 102. Multiple exposure attachment
with inserted mask 24×36 mm.

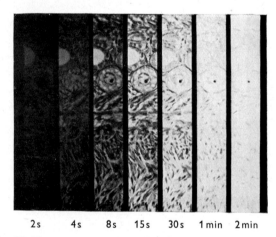

2s   4s   8s   15s   30s   1min   2min

Fig. 103. Test exposure made with multiple exposure attachment.

Test exposures are also important in colour work, particularly with regard to the lesser exposure latitude. However, expensive colour film need not be used for the test exposure; a hard-working panchromatic black-and-white film such as Agfa Isopan FF or its Kodak or Ilford equivalent is sufficient. Assessment of optimum exposure time is made easier by subjecting the comparison film to reversal development[26]; but as this is rather time-consuming (about 30 minutes), most workers prefer judging the black-and-white negative.

The initial technique is as follows: An exposure series is made on black-and-white film of a coloured test specimen which should include all three primary colours (exposure increase factor $2 \times$). Then a similar series is made on colour reversal film (exposure increase factor $2 \times$). The best black-and-white exposure time t_S is then judged, which corresponds to the best colour exposure time t_F. The quotient t_F/t_S is then noted for the emulsion used. Changes in colour film emulsion number require a new calibration. In this way, a black-and-white test exposure series will give the corresponding exposure T_S. Multiplication by t_F/t_S gives the exposure time T_F required for the colour film. The safest method is to make three exposures ($\frac{3}{4} T_F$, T_F and $1\frac{1}{2} T_F$). One of these (sometimes even two) shows optimum colour rendering. Disadvantages are loss of time, extra use of black-and-white film or plates, and the necessary two separate cameras for rational work. The fact that in addition to the colour photograph, a black-and-white picture is also obtained, may be counted as an advantage.

6.162 Exposure calculation after making a test exposure. Let T be the exposure time for a given objective-eyepiece combination, where the numerical aperture of the objective be $N.A.$ and the total scale of reproduction M, then the following equation will be valid for other combinations:

$$T_x = T \cdot \frac{N.A.^2 \cdot M_x^2}{N.A._x^2 \cdot M^2} \tag{35}$$

provided that illumination aperture and objective aperture have the same relationship each time and that the illumination of the aperture diaphragm of the microscope can be considered identical in both cases (see 3.114). The specimen may be exchanged for another of approximately the same absorbing or reflecting power. This calculation yields sufficiently blackened black-and-white negatives within the given limits, but is not recommended for colour work because of the lesser exposure latitude.

6.163 Exposure meters. Subjective or objective measurement of the exposure time is always superior to the above described methods in the aspect of gain in time.

The simplest instruments for measuring the exposure time are optical exposure meters, placed at the exit pupil of the microscope. The meter is provided with numbered fields of different density and the reading is determined by the object detail which is just discernible through one of the fields. The adaptation of the eye has to be kept constant, which can be achieved by making each measurement after perhaps 5 seconds[25]. This method is adequate for black-and-white work, but not for colour.

A photoelectric pocket exposure meter gives more accurate readings. It is used by placing it in a home-made mount on the micro adapter, or on the attachment camera, so that the distance between meter and microscope remains constant and the values are reproducible. Characteristic of this use of a pocket exposure meter is the initial sensitivity, i.e. the luminous intensity required to produce the first measurable deviation of the pointer (approximately 1 mm pointer displacement). Depending on the meter being used, this minimum value is not often lower than 1.5 lux. A comparison with the luminous intensity of hundreds of miniature film photographs, read in the film plane (Table 12) convinced us that the photoelectric exposure meter gives only good results with miniature photomicrography by bright-field transmitted light, and poorer results with photography of rock sections by polarized light.

The exposure meter is calibrated by relating the optimum exposure time from a series of test exposures, to the exposure time shown by the meter when set at the correct film speed. This provides the value of the lens aperture, at which all readings most be done. Calibration obviously depends on subjectively varying development.

Special exposure meters have been designed for photomicrography, consisting of a photoelectric detector and corresponding measurement instrument. Such a device should give an immediate reading of the exposure time. Calibration, aperture and magnification factors should not be taken into consideration, as these complicate unnecessarily the measurement. The exposure time thus obtained should be the optimum value. Absolutely correct exposure is of paramount importance since colour film reacts to small changes in exposure time by considerable shifting of colour values. Selenium and cadmium sulphide (CdS) photo-cells are most frequently used. Their colour sensitivity lies approximately in the centre of that of orthochromatic and panchromatic emulsions, with the exception of the long-wave infra-red band. Since here is no great deviation towards either end of the spectrum, a good correspondence to the colour sensitivity of the film can be assumed. This is also valid for colour emulsions. The selenium photo-cell is not expensive; it takes up little space, and does not require an additional source of energy. The CdS photo-cell does however require the latter. Exact measurement can only be guaranteed when the detector is arranged in the film plane or one of its conjugate planes. It should be remembered that measurement of image intensity only occurs over an area circumscribed by the image format. Only then can this method be used with certainty for dark-ground, polarized and fluorescence photomicrography. When these two conditions are not fulfilled (as in the case of the later described measuring device for bellows cameras) and the total beam emerging from the projection system is measured, the increased photo-current is an advantage. The measurement device can also be of lesser sensitivity. The measured exposure time is sufficient for black-and-white emulsions for densities within the exposure latitude, while in the case of colour film it determines the approximate value of the exposure. In the latter instance it is therefore recommended to make three slightly different exposures. Practice has shown that despite the integrating principle of measurement, specimens are correctly exposed which have only a few intensely illuminated details on a dark background (dark-ground, polarized light), or a few intensely coloured particles on a bright background. Fig. 104 shows part of a miniature film with consistently correct exposures, measured with a photoelectric exposure meter. The choice of meter depends on the applications: when photomicrography is restricted to photography in bright-field and phase contrast by transmitted light, a meter with a measurement range of about $5 \cdot 10^{-8}$ to $5 \cdot 10^{-5}$ ampere will be sufficient. For polarization, darkground and incident light photomicrography, the measurement range should be much greater.

Calibration is effected as follows: the photo-cell is mounted in a light-proof mount and coupled interchangeably with the micro adapter or the attachment camera, so that the photosensitive area of the cell lies in the image plane. The illumination intensity in the film plane is varied by means of a rheostat, and an exposure series with the usual steps (preferably four or five) is made at each of five, six or seven points of measurement, which are uniformly spread over the total measurement range of the instrument (e.g. $a_1 = 100$ scale graduations, $a_2 = 50$, $a_3 = 25$, etc.). The test object should have as little structure as possible; a simple centring cross has proved satisfactory. After processing and drying of the film, the various densities are measured. This can be done by relating the densities of the test exposures to an unexposed area of the film or plate (fog), with the microscope with photo-cell and galvanometer. The deviation a_0 of the measurement instrument is then adjusted (by means of the rheostat) to value 100 for an unexposed area. The densities of the test exposures will then give deviations a_1, a_2, a_3, etc., from which the densities can be calculated:

$$D = \log \frac{a_o}{a_v} = 2 - \log a_v$$

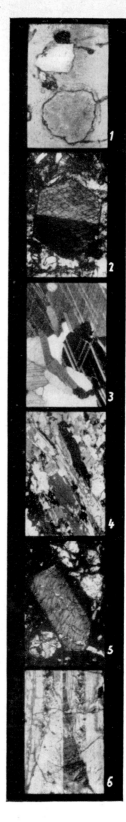

Fig. 104. Part of a miniature film; the exposure times were determined with an exposure meter.

from top to bottom:

Pitchstone porphyry 1 polarizer; $M = 20:1$ 8 seconds.

Monchiquit, crossed polarizers; $M = 50:1$ 10 minutes.

Marble, crossed polarizers; $M = 16:1$ 40 seconds.

Amphibolite, crossed polarizers; $M = 25:1$ 1 second.

Limburgit, crossed polarizers; $M = 20:1$ 95 seconds.

Limburgit, crossed polarizers; $M = 100:1$ 4 minutes.

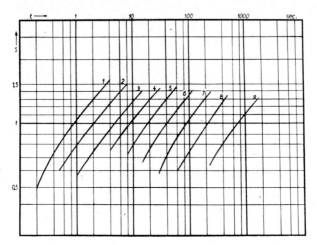

Fig. 105. Graphical representation of the density values of 9 test strips (Table 13) against the exposure time. Each point of intersection of a curve with the horizontal line $S = 1$ shows the required exposure time for galvanometer deviations a_1, a_2, etc.

The calibration value each time is the exposure needed to produce a density $D = 1$ above fog. These values can be plotted in relation to the density of the test exposure (Fig. 105). When the exposure times obtained at the measurement points, belonging to $D = 1$, are plotted against the galvanometer deviation a_v, a calibration curve is produced for the film tested, from which the exposure time can be read off directly (Fig. 106). Table 13 is an example of such a calibration.

According to the Bunsen-Roscoe law, the calibration curves must be straight lines in the chosen arrangement, showing an angle of $-45°$ with the ordinate. In practice, the angle of inclination is hardly ever exactly $45°$, and is different for each photographic material. The cause of this deviation is reciprocity failure, which is already perceptible in the exposure range chosen here. The blackening is mainly not proportional to the exposure (light intensity × time), or $E = I \cdot t$, but requires the inclusion of a factor p to qualify the exposure time: $E = I\,t^p$. It has, however, been found that this factor p is variable, and depends on t and E. The tangent of inclination of our calibration straight lines indicates the mean value of p for the material under consideration.

The method described above can only be employed rationally by using a miniature or medium-format camera in conjunction with the camera attachment or a similar device. The measurement of the exposure time on the film plane of a large-format camera (bellows cameras 9×12 cm, 13×18 cm, 4×5 in.) would necessitate a far more sensitive instrument because of the lesser illumination intensity. The following method can be used with a selenium photo-cell[28]. The cell is arranged (by means of a special arm) at the distance above the microscope whereby the photosensitive area is just illuminated by the eyepiece with the lowest power likely to be used. The galvanometer deviation obtained in this position is then calibrated to the exposure times

124

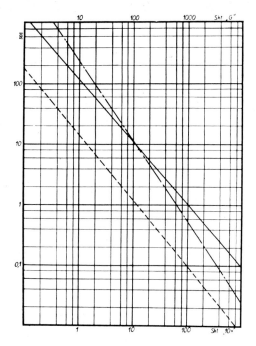

Fig. 106. Calibration curve for different film (exposure determined by means of photo-cell and double-range galvanometer).

——————— Isopan FF, Atomal F, 4 minutes at 20 °C.

— — — — Isopan F, Atomal F, 8 minutes at 20 °C

—·——·——·— Agfacolor UT 16.

required for a camera extension of 250 mm ($M_{\text{picture}} = Mg_{\text{microscope}}$). When changing the extension to k mm, the exposure read off the calibration curve must be multiplied by:

$$F = \left(\frac{k}{250}\right)^2 = \left(\frac{M}{Mg}\right)^2$$

where M is the scale of reproduction at k mm bellows extension, and Mg the total magnification of the microscope.

This method is not as exact as measurement in the image plane, since marginal parts of the eyepiece image field, which contribute to the measurement but not to the image photographed, can cause errors in the exposure time. These deviations lie within the exposure latitude of the material. The described measurement is only suitable for work with the microscope but, with only a small modification, can also be used for macrophotography. As the position of the photoelectric detector has been adapted to the possibilities of the microscope, and is usually close above the camera shutter, in macrophotography the detector comes very close to the macro-lens. The luminous intensity on the photo-cell is thus considerably higher than with the multi-stage imaging in the microscope. The calibration curve found for the microscope cannot be used for macro-work. Practice has shown that the exposure values of this calibration curve can be multiplied by a constant factor (e.g. 0.2 or 0.3). The reduced macro calibration curve thus obtained is then calibrated for any photomicrographic lens in the following way: previously carried out manipulation determines the relationship between galvanometer and exposure. For a given reading, the scale of reproduction is established by varying the camera extension, which produces, in conjunction with the exposure time obtained from the reduced calibration curve, density of $0.8-1.0$ above fog. The reduced calibration curve is then valid for this defined scale of reproduction (M_{cal}). For any other scale of reproduction with the same lens, the exposure time obtained from the calibration curve must be multiplied by the factor $(M/M_{cal})^2$.

When a selenium photo-cell is used, the value of the exposure time given by the calibration curve is also valid when a filter is inserted in the illumination beam. A red filter is an exception, however. Here, the exposure time is measured with unfiltered light, and the value obtained multiplied by the filter factor of the red filter. When a xenon high-pressure lamp is used for exposure measurement with selenium photocells, it is always necessary to insert a heat-absorbing filter, which should also remain there during photography. Measurements without heat filter would falsify the exposure time by the factor 5, because of the strong red and infra-red emission of the xenon high-pressure lamp, since the selenium cell, and not the film, reacts to this radiation.

6.164 Automatic exposure meters. Automatic exposure meters used in photomicrography are so designed that the photo-current of a photoelectric detector (most often photo-cells, sometimes also photo-multipliers) is used for the control of a special shutter. When the automatic device can be combined with commercially available camera attachments, the shutter built into the latter will not be required. Therefore, it is set to "*T*" before an exposure is made. When there is no automatic device, camera attachments where shutter and film transport are not coupled together are best suited for automatic exposure control.

The basic function of an automatic exposure device is as follows: an electric pulse causes the special shutter to open; the image-forming pencils of rays now impinge simultaneously on the photographic material and the photoelectric detector. The latter's photo-current charges a capacitor of given capacitance. When the capacitor is fully charged, the shutter is closed by means of a relay.

It follows that charging time and exposure are in direct relation to each other. Since photographic materials of different speeds need different exposures by the same illumination conditions, it follows that different capacitors with graduated capacitance must be available. These capacitors can be calibrated definitely for different film speeds. We obtain automatic exposures by the various switching positions which can be adjusted on the automatic device, by keeping constant the luminous intensity in the film plane.

126

With the negatives thus obtained, the densitometer allows us to determine the switch position which gives a density = 1. In the case of colour film, a more subjective method must be used to determine the exposure which gives optimum colour rendering. Note, however, that the chosen luminous intensity should not be too high for this calibration. A good value is 2—3 lux, which corresponds to an exposure of 4—6 seconds on 8 ASA film. If a higher luminous intensity is used, there is a risk of reciprocity failure in the case of very short or very long exposures since the automatic device does not generally take this effect into consideration, and can only be made to do so at considerable expense.

With large-format cameras with variable extension, the value thus obtained is only valid for a given camera extension. It is recommended to carry out this calibration at a camera extension of 250 mm. For other extensions, the exposure has to be amended according to a table supplied by the camera manufacturer. Apparatus with the detector in the film plane are, of course, an exception to this rule.

When filters are used, the calibration values must be corrected. This depends on the design of the automatic device and the respective values can usually be found in the instructions for use.

It is useful to make a comparison between automatic exposure and exposure measurement devices, although any comparison would be open to a great many subjective considerations. Automation is superior because of the omission of a number of operations (reading the meter, obtaining the exposure time from the calibration curve, setting the shutter). On the other hand, when calibrating a measuring device, the reciprocity failure is automatically taken into account, thus avoiding possible failures. The opportunity of selecting the measuring range of an exposure measuring device by suitable meters and adapting it to the given conditions can be regarded as points in favour of the measuring device (particularly from a financial point of view). Summing up, we can therefore say that both methods have their advantages, and neither can outweigh the merits of the other. Automatic exposure is preferred for routine work in bright-field and phase contrast photomicrography. The exposure measuring method is more advantageous where extreme variations exist in the exposure times, particularly in photomicrography by polarized light.

6.165 Illumination regulators. It is often required to regulate the illumination intensity in the image plane continuously and measurably, without affecting the image character and resolving power. For instance, this is necessary when the measured exposure time does not tally with the camera shutter speed, but when exact exposure is required for the restricted exposure latitude of the colour film. A similar regulation of the illumination is also necessary with electronic flash work. The aperture diaphragm of the microscope should never be considered as an illumination regulator. Neither can the operational data of the light source be changed, since this causes alterations in colour temperature.

The simplest method is using a set of neutral density filters of varying density.

They influence neither the colour temperature, nor the picture quality. A disadvantage is that their density is not infinitely variable.

Crossed polarizers are frequently used for this purpose[29], which allows the light intensity to be varied between a maximum value and practically zero. For this type of illumination regulation the following equation is valid:

$$J = f \cdot J_0 \cdot \cos^2 \varphi$$

where J = emerging and J_0 = incident intensity; φ = angle between the two directions of vibration. With the usual polarizing filters, f is between 0·10 and 0·25, so that even at maximum adjustment a loss of light of between 75 and 90% has already occurred. A further disadvantage is that regulation is non-linear, and occurs as a $\cos^2$ function, so that in the end position, rotation through a narrow angle results in a great change of intensity.

Both drawbacks are lessened by using neutral wedge filters. The maximum transmission can be as great as 80%, regulation is logarithmic:

$$J = J_0 \cdot 10^{-K \cdot \varphi}$$

where φ is the central angle, and K the wedge constant.

Two opposite wedges are required in order to ensure uniform light reduction over the entire range. But their absorption is frequently dependent on wavelength, while the wavelength also varies with density. In black-and-white photography, these two drawbacks are barely noticeable. But they render exact exposure measurement impossible in colour work, producing a slight colour cast even at maximum transmission. Polarizing filters are preferred for use here since they do not have these defects.

Illumination regulators should in principle be inserted in the illuminating beam.

6.2 Photomicrography by transmitted light with the microscope

Transmitted light is used for the photography of transparent specimens. At scales of reproduction of 40 : 1 and larger, the two-stage reproduction method is employed in conjunction with the microscope. The type of equipment used depends on the specimen and on the points discussed in the previous sections. The specimen is first focused visually with the eyepiece (without the camera) and the exact adjustment of the illumination is carried out. The camera is then attached to the microscope, followed by final focusing.

6.21 Bright field

6.211 Köhler's method. The most appropriate microscope lamp is arranged at the correct distance from the microscope and securely attached to the microscope. The field diaphragm is imaged by the substage condenser in the objective's object plane. First, a low-power objective (6—8×) and eyepiece are used to focus the microscope on the object plane, without considering the illumination. All that is necessary here is that the specimen is sufficiently illuminated for its contours to be discernable (Fig. 107a). The field diaphragm is then closed down to a small value and the substage

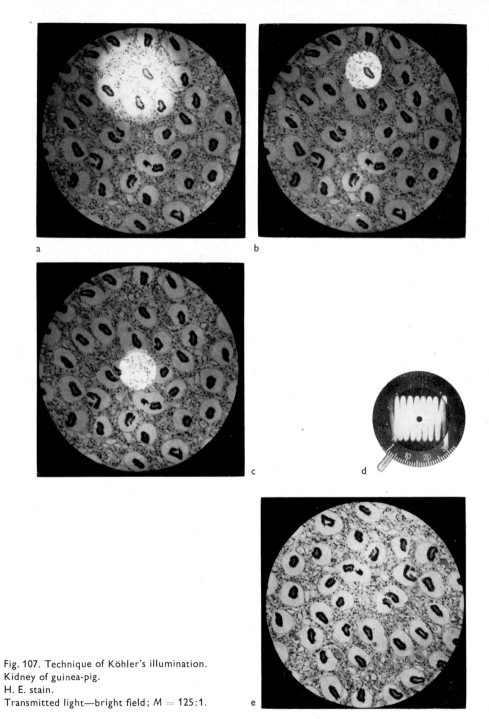

Fig. 107. Technique of Köhler's illumination.
Kidney of guinea-pig.
H. E. stain.
Transmitted light—bright field; $M = 125:1$.

condenser is raised or lowered until an image of the aperture of optimum sharpness is obtained (Fig. 107b). The illuminating mirror is then used to bring the image into the centre of the field (Fig. 107c). The next step is to focus the light source so that a sharp image of the filament is formed on the closed aperture diaphragm (Fig. 107d). This light source image should fill the largest possible aperture of the aperture diaphragm and be well centred. If the image is smaller, the light spot which appears after insertion of a ground-glass screen should at least fill the entire aperture of the diaphragm. The field diaphragm is now opened so that the eyepiece object field is just completely illuminated (Fig. 107e). If at the chosen objective-eyepiece combination it is not possible to illuminate the eyepiece object field completely, even at full opening, the focal length of the inserted condenser must be increased by removing the top lens of the condenser. This may be done either by unscrewing, or by swinging out (see 3.221). The same manipulation is necessary when it is not possible to reduce the image of the aperture diaphragm in the exit pupil of the objective to a sufficient degree by stopping down the diaphragm.

The aperture diaphragm is adjusted in accordance with the structure of the specimen. It is the purpose of the aperture diaphragm to obtain the best possible compromise between optimum resolution and the degree of contrast required for reproduction. A too large aperture results in weak images of insufficient contrast, while a too small aperture gives strong diffraction at the edges of the specimen (see Fig. 120). Any dust particles situated in the vicinity of the specimen or its image on lens surfaces, or on the object carrier, will also become visible. A simple control is offered by looking at the diaphragm image in the exit pupil of the objective, which is easily seen after removing the eyepiece. At a suitable position of the diaphragm, the image will usually fill one to two thirds of the area of the exit pupil of the objective.

When working with objectives of a numerical aperture 0·65 and higher, one can observe a decrease in sharpness of the image of the field diaphragm in the object plane, when opening up the aperture diaphragm. Neither is it easy fully to illuminate the exit pupil of the objective, although the numerical aperture of the condenser is completely adequate for this purpose. The cause of both phenomena is found in the lack of spherical correction of the condenser. This can be remedied by moving the condenser to a position whereby the exit pupil of the objective is fully illuminated at a corresponding value of the aperture diaphragm. When working with immersion objectives, immersion of the condenser is required (see 3.221). In this way, aplanatic and aplanatic-chromatic condensers are also considerably improved in spherical correction and their focal intercept is increased.

6.212 Simple illumination. When no microscope lamp is available which is suitable for Köhler's illumination, simple illumination must be used, usually consisting of only a tungsten lamp and a mirror. As the coil of the tungsten lamp will usually cover too large an area, a groundglass or opal-glass diffusing screen must be used. A point to watch is that the grain of the ground-glass screen should not be visible in the object plane. The substage condenser should therefore be moved up or down until a uniform, grain-free illumination is obtained. It is obvious that the coil of the lamp should be placed at such a distance from the diffusion screen that no uneven illumination is produced on the screen. The latter should be made as homogeneous as possible.

This simple illumination system should only be used for photomicrography in an emergency.

6.22 Dark ground

6.221 Illumination by dry condenser. Dry condensers are usual in conjunction with objectives whose aperture does not exceed 0·8, i.e. condensers without immersion fluid between top lens and underside of slide. Dark-ground illumination can be obtained with bright-field condensers fitted with a diaphragm. But special mirror condensers are to be preferred to such makeshift arrangements (see 3.222). They have superior spherical correction and produce a better dark ground (Fig. 108). The universal condenser illustrated in Fig. 60 is a good example of a dry dark-ground condenser, with its focal intercept of 11 mm and N.A. 0·8. This condenser allows a rapid change from dark-ground to brightfield illumination by swinging out of the diaphragm (N.A. 0.5). The condenser should be well illuminated by a correct central arrangement of the microscope lamp. After having swung out the dark-ground diaphragm, an exact adjustment of Köhler's illumination is carried out. The diaphragm is then swung back, the aperture diaphragm opened fully, and the condenser moved up and down

Fig. 108. Magnesium ammonium phosphate. Transmitted light-dark ground; $M = 100:1$.

until the narrowest part of the light cone emerging from the condenser lies in the object field. To obtain perfect azimuth-free dark-ground illumination, it is often necessary to re-adjust the mirror position. Insufficient adjustment is made apparent by unilaterally illuminated parts of the specimen.

6.222 Illumination by immersion condenser. High-aperture objectives require immersion dark-ground condensers. The space between the top lens of the substage condenser and the underside of the slide must always be filled with an immersion fluid; otherwise object illumination would be impossible because of total reflection of the light rays by the slide. The numerical aperture of an objective used with an immersion condenser should be smaller than the minimum aperture of the condenser, otherwise bright-field illumination would result. High-aperture objectives for dark-ground work are therefore fitted with an iris diaphragm.

An exceptionally good union of the illuminating rays in the object plane is most important. This condition, combined with good spherical and chromatic correction, is best fulfilled by a mirror condenser of special design, the cardioid condenser.

To obtain an azimuth-free overall dark-ground illumination, the centre of the union of illuminating rays should be perfectly centred in the microscope's optical axis. A cardioid condenser should therefore always be fitted with a centring device.

The cardioid condenser is inserted in the illumination device of the microscope, instead of the bright-field condenser. Any stage diaphragm should be removed, and a spot of immersion oil be dropped on the top lens of the condenser. The slide is then placed on the stage, and the illumination device with condenser moved upwards until the fluid touches the slide and spreads over its surface, without any air bubbles, of course. Should the latter occur, careful sliding of the specimen will remove the air bubbles from the object field. The slides should have a thickness smaller than 1·2 mm, corresponding to the focal intercept of the condenser. Since the illuminating rays must intercept each other in the object plane, slides of greater

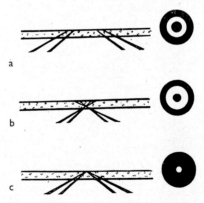

Fig. 109. Adjusting the cardioid condenser; *a* Wrong! condenser too high, *b* Wrong! condenser too low, *c* Correct adjustment.

thickness cannot be used. On the other hand, slides which are too thin may cause the immersion fluid to be removed, because of the required lowering of the condenser.

After having adjusted the microscope lamp so that parallel light fills the centre of the cardioid condenser (via the centre of the illuminating mirror), the object is focused by means of a low-power objective and eyepiece. The condenser is then raised or lowered until the light spot seen through the eyepiece has reached its narrowest position (Fig. 109). This light spot is then brought into the centre of the field by means of the centring device. After switching to the objective which will be used for photography, the condenser may be re-adjusted if necessary. These adjustments must be carried out with great precision, if good dark-ground results are to be obtained.

6.3 Macrophotography by transmitted light

One-stage reproduction is to be preferred for scales of reproduction of up to 40 : 1. One of the systems described in 3·321 can be used as objective. The choice depends on the size of the specimen and the required scale of reproduction (Table 14).

The lenses can be used with special large-field condensers (spectacle glass condensers), specially selected for their free aperture and back focus. They should be arranged as close to the object as possible. They have the task of illuminating the specimen evenly and imaging the light source in the entrance pupil of the lens.

Since the use of photomicrographic lenses usually excludes visual observation, focusing and subsequent adjustment of the illumination device is carried out on the

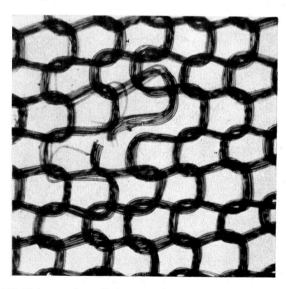

Fig. 110. Nylon stocking. Transmitted light—bright field; M = 32:1.

133

focusing screen of the camera. A miniature camera or camera attachment should be directly coupled with the microscope stand. Camera attachments with focusing systems cannot always be used for macrophotography, since their design causes a strong vignetting of the image rays emerging from the photo-objective. The same can happen with a normal microscope tube with an inside diameter of 23·2 mm. The best stands for macrophotography are those which allow the camera to be arranged at a very short distance above the objective changing device.

6.31 Bright field

6.311 Köhler's illumination. When Köhler's illumination is necessary for retaining unused object areas in macrophotography, the illumination arrangement should be carried out as shown in Fig. 111. A lens (*6*) is inserted between spectacle glass condenser (*7*), which is situated close to object plane (*8*), and bull's-eye condenser (*2*), which

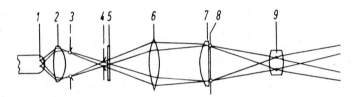

Fig. 111. Köhler's illumination for macrophotography (transmitted light).

serves to image the field diaphragm (*3*) in the object plane by changing its distance from the condenser. Light source (*1*) is imaged by the bull's-eye condenser on the aperture diaphragm (*4*) and the latter with the auxiliary lens and the spectacle glass condenser in the entrance pupil of the objective (*9*). The focal length of the auxiliary lens should be such that the image of the light source produced by the bull's-eye condenser, the auxiliary lens and condenser in the entrance pupil of the objective, will just illuminate equally. In case the filament structure should become apparent, it is best to insert a finely ground glass screen between tungsten lamp and bull's-eye condenser. A filter (*5*) can be inserted in the vicinity of the aperture diaphragm.

6.312 Simplified arrangement. Köhler's illumination is rarely used for macrophotographic reproduction, since it requires optical equipment of a high standard; and an equally high picture quality can also be obtained with the macrophotographic camera using simpler systems of illumination (Fig. 112). The latter is illustrated in Fig. 45, where a light source (*1*) of the microscope lamp is imaged at infinity by the bull's-eye condenser (*2*). A spectacle glass condenser (*3*) of corresponding focal length is arranged closely beneath the object plane (4), and images the light source on the aperture diaphragm of objective (5).

The arrangement described in 6·212 can also be used for illumination, but care should be taken to obtain a good, uniform illumination, because of the larger objective object field.

6.32 Dark ground. A simple arrangement for dark-ground illumination with one-stage reproduction, can be obtained by inserting a central diaphragm in the illuminating beam between bull's-eye condenser and substage condenser. With the simplified illumination system of Fig. 45, this diaphragm's position would be in the proximity of the bull's-eye condenser. When Köhler's illumination is used, the diaphragm would be positioned closely beneath the substage condenser. The diameter of the central diaphragm can be determined empirically, so that a tube-shaped illumination beam fills the rear aperture of the spectacle glass condenser. The iris diaphragm of the

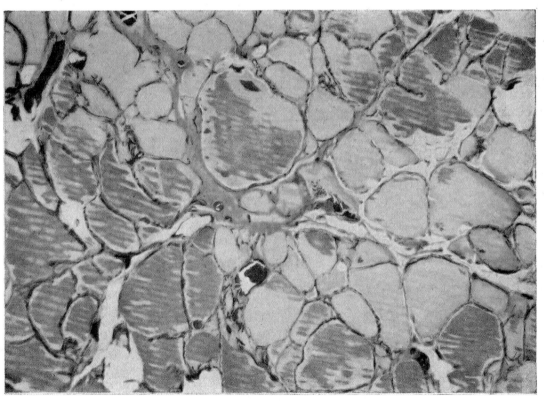

Fig. 112. Struma; colloid Goldner staining. Transmitted light—bright field; M = 10:1 in original enlarged to M = 40:1 Agfacolor UK 16.

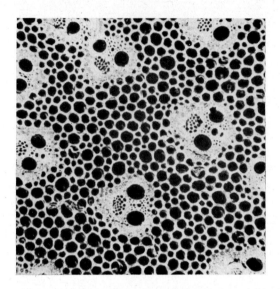

Fig. 113. Bamboo, section. Transmitted light—dark ground; $M = 50:1$.

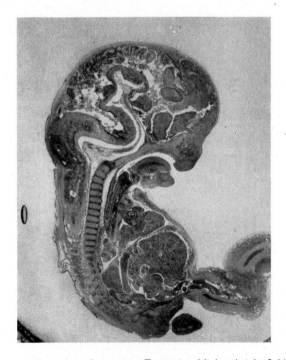

Fig. 114. Human embryo, 6 weeks; eosine. Transmitted light—bright field; $M = 3,2:1$.

objective is so closed that no direct light can be seen any longer, when observing the exit pupil. The described arrangement produces an overall dark-ground illumination (Fig. 113). Oblique illumination can be obtained by an appropriate shape of the central diaphragm.

6.4 Low-power photomicrography by transmitted light

With photography at scales of reproduction of 1 : 1 to about 4 : 1, the objective object field is so large that special spectacle glass condensers with larger diameter and longer focal length are required for illumination. The object is reproduced by means of an objective whose long focal length makes it impossible to fit it to a normal microscope stand; it must be directly coupled with the camera (see Table 15). The same applies in principle to macrophotography where $M < 1 : 1$.

6.41 Bright field. A special arrangement is required for photographing large objects by bright-field illumination, because not only must the object field be well illuminated, but this light must also reach the objective used for photography. If possible, the light source should be imaged on the aperture diaphragm of the objective. This is achieved by inserting a large spectacle glass condenser underneath the stage, so that the object is very close to the surface of the condenser. In order to illuminate fully this condenser, a diverging lens is inserted between bull's-eye condenser and

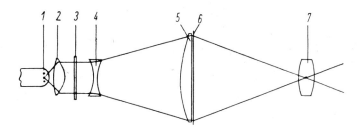

Fig. 115. Arrangement for low-power transmitted light photography.

spectacle glass condenser. Fig. 115 shows the path of rays with this arrangement. (*1*) is the light source, (*2*) the bull's-eye condenser, (*3*) a finely ground glass screen, (*4*) the diverging lens, (*5*) the spectacle glass condenser, (*6*) the object plane, and (*7*) the objective.

6.42 Dark ground. In order to make low-power photographs by transmitted light dark-ground illumination, (Fig. 116), a few changes are required in the illumination system. A simple method is to conduct the light by means of a mirror through the object to one side of the objective, without any direct light entering it (Fig. 117).

137

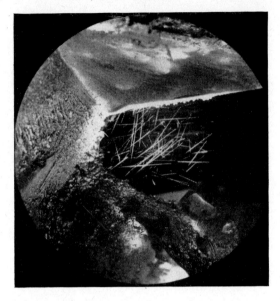

Fig. 116. Quartz crystal with acicular incasements. Transmitted light—dark-ground (oblique); $M = 16:1$.

The light source (*1*) can be displaced against the bull's-eye condenser (*2*) so that a parallel beam of light falls on the mirror (*5*). Light diffracted in the specimen (*3*) enters objective (*4*) and contributes to the image formation. This method obviously gives unilateral dark-ground illumination. By encircling the microscope with several microscope lamps with mirrors, a multilateral, almost overall, dark-ground illumination can be achieved.

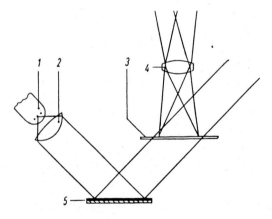

Fig. 117. Arrangement for low-power unilateral dark-ground photography.

Both bright-field and dark-ground illumination are used for the photography of opaque specimens. Bright-field illumination is primarily used with flat specimens, such as polished metal or ore sections which give regular reflection and whose structure can be made visible by suitable etching methods. Details, where the light is absorbed or diffracted, appear darker in the print than regular reflecting surfaces.

Dark-ground illumination, on the other hand, is most suitable for revealing very fine detail on regular reflecting surfaces, which would be invisible in bright -field illumination when they are smaller than the resolving power of the lens used. This is also the best illumination for colour rendering and representation of detail below a transparent lacquer, glazing, etc. The photographing of specimens with a pronounced three-dimensional surface, especially at medium or low-power magnification, is a main field of application.

Additional devices for incident light illumination are variously known as vertical illuminators, reflectors, or epi-objectives. Special incident light microscopes have been designed using *Le Chatelier's* system of arranging the illumination device beneath the stage (Fig. 95). The advantage is that opaque objects can be placed directly on the stage, which is perpendicular to the microscope axis. In this case, there are no restrictions on the thickness of the object. In accordance with the properties of vertical illuminators, those objectives which are corrected for an infinite image distance are preferred. A tube system in the microscope or in an additional tube ensures that the image of the object is formed on the eyepiece object plane.

6.51 Bright field. Bright-field illuminators for the compound microscope are always arranged so that there is a reflecting surface in front of the objective, which may be a cover glass, plane-parallel glass or prism. The light of a microscope lamp arranged at the side of the instrument reaches the object through the objective, which in this case acts as condenser. When illuminators are used which allow a change between prism and glass, it should be remembered that plane-parallel glass slightly reduces the quantity of transmitted light, but in no way changes the numerical aperture of the objective. The prism, on the other hand, covers half the objective aperture on the image side. The numerical aperture of the objective in use (and hence the resolving power) here only remains unchanged in the direction parallel to the edge of the prism. Fig. 118 shows the effect of the type of reflector on resolving power. The use of a prism also makes difficulties in illuminating the entire field; therefore the tendency is towards the use of plane-parallel glass.

The only advantage of the prism for bright-field illumination is the greater brightness of the object. But this advantage is now outweighed by the use of specially coated glasses and high-power light sources. Only for polarized light work should a prism be used; namely, a special compensating prism, because a plane-parallel glass would depolarize the light and hence make it more difficult to diagnose anisotropic objects.

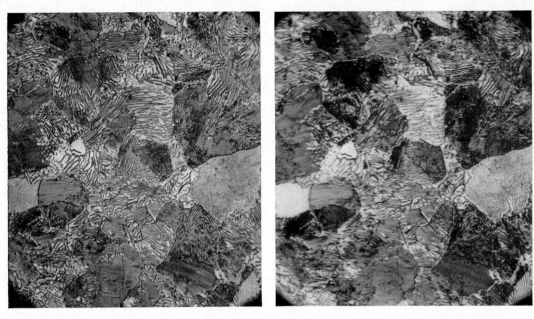

Fig. 118. Comparison between incident light—bright field illumination with plane-parallel glass (left) and prism (right).
Steel with 0.9% C alcoholic nitric acid; M = 630:1.

6.511 Köhler's illumination. This type of illumination requires a different arrangement for incident light to that for transmitted light. Fig. 119 shows the arrangement for incident light, where light source (*1*) is imaged on aperture diaphragm (*3*) through a bull's-eye condenser (*2*) at such a size that the maximum aperture of this diaphragm is entirely filled. A lens system (*4*) and (*6*) images the aperture diaphragm through the plane-parallel glass or prism (*8*) in the rear focal plane of objective (*9*). The field diaphragm is imaged by lens (*6*) at infinity, and by the objective in the object plane

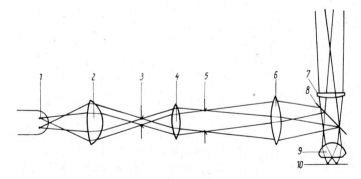

Fig. 119. Köhler's illumination for the compound microscope (incident light).

(*10*). The tube system (*7*) ensures the imaging of the image produced by the objective at infinity, in the eyepiece object plane. If the position of auxiliary lens (*6*) to the field diaphragm is so arranged that the diaphragm lies in the focus of the lens, the field diaphragm image will always be formed in the objective field plane, provided that the objective has an infinite image distance. This can be used to find this plane. High-power objectives have little depth of field, which can make it difficult to find the focusing plane, particularly in the case of objects which have little or no contrast. The objective can easily come into contact with the specimen, with possible damage to both. This can be avoided by focusing with a narrowly closed field diaphragm for as long as is necessary to obtain a sharp image of the field diaphragm. This image lies in the object plane, and is far easier to establish than the surface of the specimen itself. The field diaphragm is opened up just far enough to completely fill the image field as

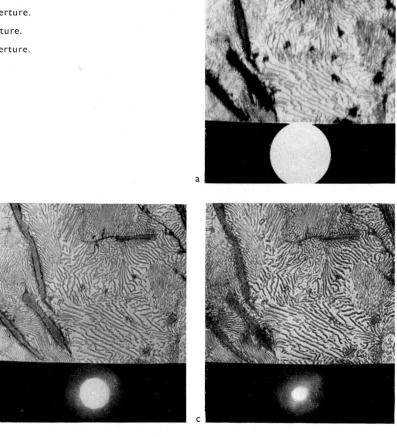

Fig. 120. Influence of aperture diaphragm value on picture quality. Gray cast iron—alcoholic nitric acid Incident light—bright field; $M = 630:1$.

a too large aperture.

b correct aperture.

c too small aperture.

in the case of transmitted light. With diffusely reflecting objects, it is often advantageous to close the field aperture still further (see 8.11).

The rules given in 6.211 apply to the adjustment of the aperture diaphragm in bright-field illumination. Fig. 120 illustrates the effect of the aperture diaphragm value on the quality of the photomicrograph. It is easily seen that fine detail receives too much light at a too large diaphragm aperture and is scarcely visible because of the excessive contrast, while a too small aperture produces diffraction phenomena which can falsify the photograph.

6.512 Simplified arrangement. Vertical illuminators for incident light—bright-field illumination offer a simplified arrangement, with either glass reflector or prism, or both, fitted interchangeably between objective and tube. A microscope lamp is arranged on one side at the appropriate distance, so that the bull's-eye condenser produces an image of the light source of optimum sharpness on the diaphragm fitted in the vertical illuminator. This diaphragm is here not strictly an aperture diaphragm, since noticeable differences occur in this arrangement between diaphragm position and exit pupil of the objective. The diaphragm is situated as closely to the glass or prism as feasible, in order to reduce possible vignetting to a minimum. The microscope lamp has an iris diaphragm which can be used as field diaphragm. When using objectives corrected for infinite image disstance, the vertical illuminator will comprise the necessary tube system. Fig. 121 shows this simplified illumination method, where (*1*) is the aperture diaphragm, (*2*) the prism, (*3*) the tube system, (*4*) the objective, (*5*) the object surface, and (*6*) the plane-parallel glass. A similar simplified arrangement is used for incident light condensers, where microscope lamp and vertical illuminator are combined into a single unit.

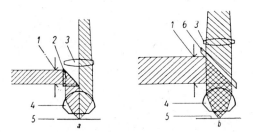

Fig. 121. Path of rays in simple incident light illuminator
a with prism, *b* with plane-parallel glass.

6.52 Dark ground. In incident light microscopy we also have illumination devices for both overall and unilateral dark-ground illumination. We make a distinction between critical and simplified illumination.

6.521 Critical illumination in dark ground. Fig. 123 shows the best arrangement for critical illumination in dark ground. Light source (*1*) is imaged on aperture dia-

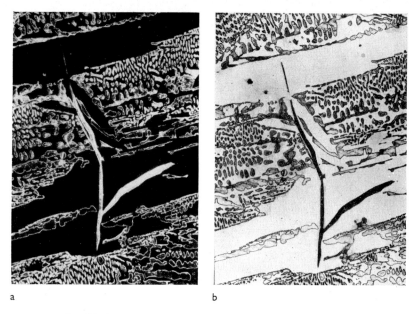

a   b

Fig. 122. Specular (cast) iron—alcoholic nitric acid; $M = 125:1$.

a incident light—dark-ground, *b* the same spot for comparison by incident light—bright field.

phragm (*4*) by bull's-eye condenser (*2*) and auxiliary lens (*3*). Lens (*5*) directs the light beam, which is now parallel, to a central stop (*7*) which absorbs the central part of the light beam. The transmitted part, which has the shape of a tube, first strikes on a 45° annular mirror (*8*), and then on a mirror condenser (*11*) which is co-ordinated with objective (*10*). This mirror condenser reflects the light to the object plane (*12*). The image of the light source is therefore formed in the object plane, thus realizing the principle of critical illumination (see 2.24). Diaphragms (*4*) and (*6*) interchange their functions as compared to bright-field work; they usually remain fully open. The auxiliary lens (*3*) should be adjusted so as to illuminate fully the aperture of the field diaphragm (*6*). The plane-parallel glass or prism, necessary for

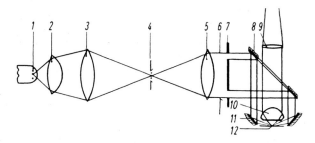

Fig. 123. Arrangement for incident light—dark ground illumination.

143

bright-field illumination, is here removed in order to avoid any reflections. The mirror condensers should be so arranged that the intersection of the emerging light rays (narrowest constriction of the light cone) coincides with the object plane when focused.

This adjustment is extremely delicate and should be carried out with great precision. Other difficulties may be caused by uneven illumination or azimuth effects.

6.522 Simplified arrangement. When for reasons of design it is necessary to forego critical illumination, we can use a considerably simpler arrangement. This type of illumination is used most frequently with incident light vertical illuminators, since it shortens considerably the overall length of the instrument. The light source is then situated in the focal point of a bull's-eye condenser. The parallel light beam emerging from the latter passes through the central stop and is directed to the concave mirror condenser by means of an annular 45° mirror. This condenser directs the light to the object plane and produces an image of the light source there. In this case, the free diameter of the bull's-eye condenser should correspond to the aperture of the concave mirror condenser, which can produce high values for the bull's-eye condenser's diameter. This is sometimes avoided by assimilating the parallel light beam emerging from the bull's-eye condenser (*2*) to the aperture of the mirror condenser (*7*) by widening the beam by means of a light trap (*3*) and (*4*), as illustrated in Fig. 124. The light trap serves at the same time as the central stop. The correct illumination of the mirror condenser requires exact focusing and centring of the light source referring to the bull's-eye condenser. When a low-voltage lamp is insufficiently centred, azimuth effects can readily occur. Insufficient focusing results in colour cast, while chromatic errors of the bull's-eye condenser also become visible.

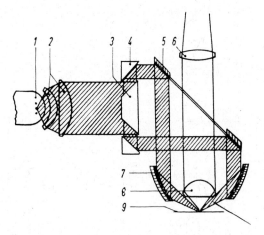

Fig. 124. Simplified arrangement for incident light—dark ground illumination.

6.523 Oblique illumination. It is easy to obtain oblique illumination from overall dark-ground illumination by removing so much from the light beam until only a sector remains. Such a diaphragm can be designed with two half-diaphragms which can rotate in opposite directions to each other, so that the azimuth can be regulated between 0° and 180°. When, furthermore, the diaphragm in the illumination space of the instrument can be rotated, the direction of the incident light beam can also be changed. In this way, the illumination can be adapted to the structure of the specimen (Fig. 125). Oblique illumination can also be obtained by arranging one or more microscope lamps next to the microscope so that their light falls from above on to the specimen. The light can strike the specimen from different directions and under different angles. The angle of incidence should be larger when the surface unevenness of the specimen is smaller. This type of dark-ground illumination can only be used when the object distance of the objective used is sufficiently long; otherwise unwanted vignetting will result.

Fig. 125. Hardness impressions with Vickers diamond pyramid in electrolytically polished and etched brass (phosphoric acid). Incident light—dark ground (oblique); M = 160:1.

6.6 Macrophotography by incident light

Photomicrographic lenses should be used in this case (3·321). Their selection should be guided by the values given in 6·3. It must be remembered, however, that the values of the obtainable scales of reproduction and object field sizes are displaced with instruments having a fixed integral tube system. These changes depend on design. It is therefore not possible to give general rules. Focusing of the specimen and adjustment of the illumination are carried out on the focusing screen of the camera as is usual with one-stage reproduction.

6.61 Bright field. Illumination is obtained by means of a plane-parallel glass, in a similar manner as in the microscope. Because of the relatively long object distance, it will usually be necessary to arrange the glass between objective and specimen (Fig. 126). The light emitted by light source (*1*) is directed to specimen (*7*) through bull's-eye condenser (*2*) and auxiliary lens (*4*), and reflected by plane-parallel glass (*5*). The focal length of the auxiliary lens (*4*) is so chosen that the light source is imaged in the aperture diaphragm of the objective (6). When changing the scale of reproduction, and hence the object distance of the objective, it will be necessary either to change the position of the auxiliary lens correspondingly, or to exchange this lens for another of a more appropriate focal length. If necessary, a finely ground screen (*3*) can be inserted in the light beam to prevent any unevenness of illumination.

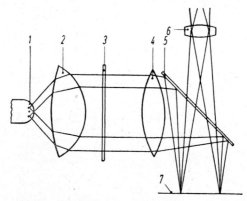

Fig. 126. Arrangement for incident light— bright field illumination in macrophotography.

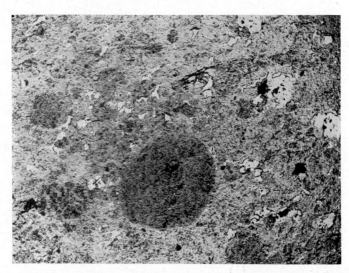

Fig. 127. Hypersthenite—chondrite (fell on the 9th June 1866 near Knyahinya [CSSR]) polished section. Incident light—bright field; $M = 5:1$.

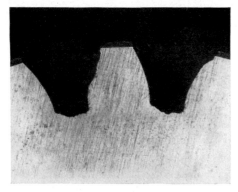

Fig. 128. Damaged tooth of a gear-wheel. Incident light—dark ground (overall); M = 4:1.

6.62 Azimuth-free dark ground. This requires the use of an annular fluorescent tube. In order to increase the light intensity, a reflector can be fitted round the tube. This method produces a darkground entirely free from azimuth effect (Fig. 128).

6.63 Oblique dark ground. The method described in 6·523, with one or more microscope lamps arranged next to the microscope, is particularly efficient for macro-photography, as no vignetting can occur because of the greater object distances of the objectives. For apparatus with a beam of light parallel to the object plane the arrangement shown in Fig. 129 is very satisfactory. The light of light source (*1*) is directed to object (*7*) through bull's-eye condenser (*2*) and auxiliary lens (*4*), by means of a mirror (*6*) arranged on one side. The direction of incident light can be varied within wide limits by tilting and shifting the mirror. This enables one to adapt the illumination to the specimen most advantageously (Fig. 130). The auxiliary lens should be illuminated over its greatest possible diameter. If its focal length is correct, iris diaphragm (*3*) can be imaged in the object plane as field diaphragm. Unwanted over-strong illumination can be reduced by means of this iris diaphragm.

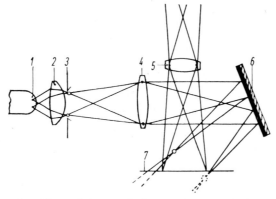

Fig. 129. Arrangement for oblique dark-ground photography in macrophotography (incident light).

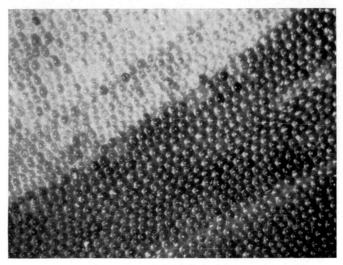

Fig. 130. Tradescantia zebrina, underside of leaf. Incident light—oblique dark-ground; angle 30°, to axis; M = 25:1.

Oblique illumination can also be achieved with an annular fluorescent tube, by cutting off the light until only a slit or sector remains for illumination. When the latter can be adjusted and rotated, excellent oblique illumination effects can be obtained.

6.7   Low-power photomicrography by incident light

6.71 Bright field. For low-power photography and bright-field illumination (Fig. 131), a plane-parallel glass, whose size corresponds to the objective diameter or object dimensions, is fitted in front of the objective. This glass should be as thin as possible and still be plane-parallel. Glass which is too thick will produce double images and astigmatic errors. Image brightness and contrast can be increased by a suitable coating of the plane-parallel glass. The light beam, which comes from one side, must illuminate the entire object field of the objective. Any microscope lamp with bull's-eye condenser and appropriate auxiliary lens can be used for this method. It should, however, be noted that, in the case of a negative auxiliary lens, the photographs may show a dark-ground effect near the edges of the image field since only light rays which are directly reflected into the objective contribute to image formation by bright-field illumination. Fig. 132 gives the explanation for the occurrence of dark-ground effects. The object is assumed to have an even surface, which is perpendicular to the optical axis of the direction of the camera. An auxiliary lens (*4*) diverges the light beam emitted by light source (*1*) after having passed through bull's-eye condenser (*2*). When the distance a—d can be taken in by objective (*6*), only the distance b—c receives bright-field illumination. The annular zone a—b and c—d will appear in dark ground, because that part of the light beam has been so reflected that it cannot enter the objective. Only rays which have been diffracted at the object surface

148

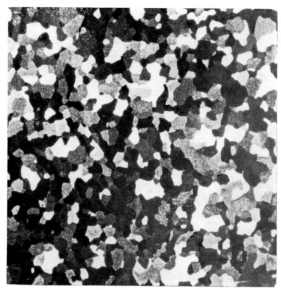

Fig. 131. Aluminium macro-etching. Incident light—bright field; M = 1:1.

contribute to the image formation of these zones. This effect can be reduced by inserting a finely ground screen (3) before the negative lens, which visibly widens the area of bright-field illumination. It is possible that a wide transition zone may occur between bright-field and dark-ground. Fig. 126 shows an arrangement for exact bright-field illumination. Here the drawback is very large lens diameters for low-power work, corresponding to the larger object field.

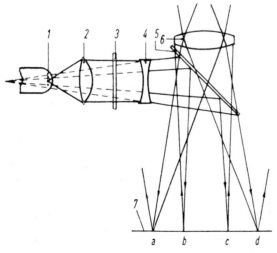

Fig. 132. Arrangement for lowpower incident light—bright field photography.

6.72 Azimuth-free dark ground. This type of illumination can be carried out with an annular fluorescent tube of sufficient diameter as already described in 6.62—even with low-power work. It is necessary to arrange the tube as closely to the object as possible. Should this distance be too great, part of the object will have a bright-field effect because of the steep incident light beam (Fig. 133).

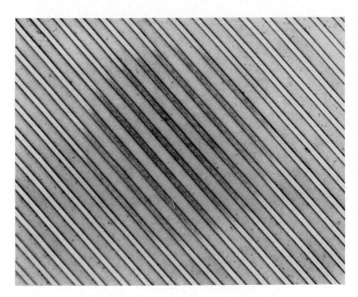

Fig. 133. Surface of a grooved stage. Incident light; transition from dark-ground (centre) to bright-field illumination (edges) caused by wrong position of the circular fluorescent tube.

Fig. 134. Steel fragment. Incident light— dark ground (oblique); M = 1:1.

150

6.73 Oblique dark ground. We can use the arrangement according to 6.63 for this type of illumination, when bull's-eye condenser and mirror are adapted to the large objective object field. With very oblique light, it is recommended to insert a slit diaphragm in the light beam in order to show up cracks and similar faults. Here, too, the annular fluorescent tube in conjunction with a laterally situated additional microscope lamp can often yield excellent results (Fig. 134). Naturally, a single microscope lamp is sufficient when it is capable of illuminating the field of the objective. In order to prevent unwanted fogging, the field should not be any larger than the imaged objective object field. Good picture quality will be more easily obtained by placing the specimen on black velvet. This also applies to the other types of illumination discussed above.

6.8 Combined transmitted and incident illumination

Some specimens must be photographed so that both the surface and the contours are clearly shown. Combined transmitted and incident illu-

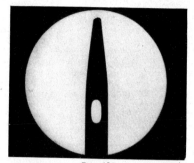
Fig. 135

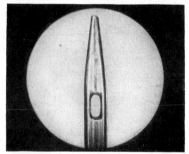
Fig. 136

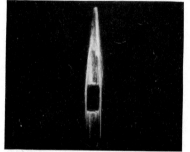
Fig. 137

Fig. 135. Sewing machine needle with broken point. Transmitted light—bright field; $M = 5:1$.
Fig. 136. As Fig. 135, but incident light—dark ground.
Fig. 137. As Fig. 135, but combined illumination.
Fig. 138. Exposure tests for Fig. 136.
Fig. 139. Exposure tests for Fig. 137.

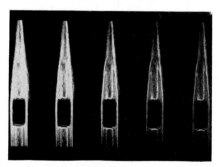
Fig. 138

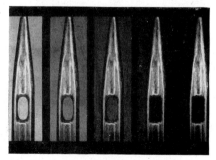
Fig. 139

mination[30] provides the solution. Fig. 135 shows the photograph of the needle of a sewing machine which has a damaged point, taken with transmitted light—bright-field illumination. Only the contours can be seen as a silhouette. The same object in incident light—dark-ground illumination provides a good rendering of surface detail, but the contours are not clearly shown (Fig. 136). A natural rendering is only obtained by combining the two illumination systems (Fig. 137). This kind of photography demands attention to the level of ground illumination. Even distribution of light is an obvious requirement. But the best tone of the background will depend on the object, and can be best determined by using a rheostat or similar device during visual observation in order to obtain the desired effect. Exposure can then take place in a single operation. Sometimes it is more advantageous to make separate bright-field and dark-ground exposures in succession. The best method is to determine the individual exposure times by means of test strips. Figs. 138 and 139 show the two ranges of test exposures from which the most suitable were selected; the combined result as in Fig. 137.

6.9 Special techniques

6.91 Macrophotography with parallel-beam (telecentric) illumination

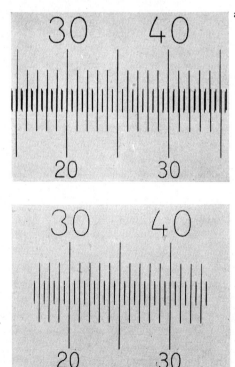

a

b

6.911 Principles. The centre of perspective will be at infinity for an object in a parallel beam of light. In the case of silhouetting an object, the size of the shadow will not change with the distance between object and projection screen; the scale of reproduction will always be 1 : 1. When an objective is placed behind an object, which should give an enlarged image or silhouette on a focusing screen, the scale of reproduction will only depend on the focal length of the objective and the image distance v (i.e. within the range of depth of field). The object distance may vary within the depth of field. Fig. 140 demonstrates this effect by a comparison of the normal arrangement with parallel-beam illumination.

Fig. 140. Two glass rulers close together:
a with ordinary arrangement;
b with telecentric illumination.

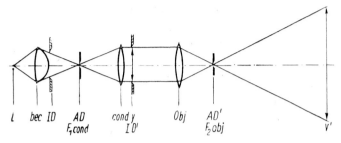

l bec ID AD $cond\ y$ Obj AD'

$\qquad\qquad\qquad F_1cond$ $I\,D'$ F_2obj v'

Fig. 141. Arrangement for photography with telecentric path of rays.

These facts make this method suitable for profile photography, such as the wear of small mechanical parts and the loss of spherical shape of ball bearings. The main field of application is technical profile projection used during production control. The optical arrangement of a profile projector is shown in Fig. 141. The light source for object y is in this case the small diaphragm AD, which is placed in the focal point of the condenser. The lens or objective images y at y' in the focal plane of the camera. Furthermore, the light source will be imaged at $F_{2\,\mathrm{obj}}(\mathrm{AD}')$. By arranging a real diaphragm at this point, it can be used for the reduction of scatter or for other purposes (see 6.92). When the light source used is sufficiently small and homogeneous, the aperture diaphragm can be arranged at this position. Otherwise a bull's-eye condenser (*bec* in Fig. 141) will be required used in conjunction with a finely ground glass screen, to image the light source at AD.

6.912 Light sources. Because the aperture diaphragm is small, the light source may also be small. Good evenness of illumination is desired; if necessary, a ground-glass screen can be used.

6.913 Filters. The choice of a filter will usually depend on the correction of the lens or objective. Green filters will therefore be used most frequently.

6.914 Optical requirements. Good photo-objectives or reversed camera lenses can be used. The objective or lens diaphragm remains open. The 'condenser' system should also be well corrected. The best method is to use an optical system similar to the objective or lens used for photography, perhaps of slightly longer focal length, but of the same relative aperture. Since the maximum useable object field corresponds to the free diameter of the camera lens (and moreover high-power work is rarely involved because of the small diaphragm aperture and the usually desired greater depth of field), objectives and camera lenses of longer focal lengths are preferred. It should be noted that the objectives and lenses mentioned above are here required to operate with a position of the diaphragm for which they have not been designed. Distortion will therefore be more pronounced than with normal use. The profile projectors used in industry are therefore fitted with systems specially calculated for parallel-beam illumination. It is obvious that mostly large-size cameras with long extensions are employed.

6.92 Schlieren photomicrography. Schlieren is the German word for streaks or striations. They are locally limited non-homogeneous areas in transparent media which cause irregular refractions of the light. Since irregularities in the surfaces of optical media have similar properties, the concept of schlieren has been extended by a number of authors to "the cause of any irregular refraction of light over a small area"[31].

Schlieren methods are optical methods for the observation of schlieren, and for the determination of their position, size and quality. Of the many existing methods we shall discuss a few which are capable of being used in photomicrography.

The simplest method is shadow projection by illumination of the object with a very small light source of great intensity. The edge of a schlieren is then darker than the free surround, since part of the light is deflected by the schlieren from its original direction to a different position on the projection screen. In this case, the image will be larger than the object. The most sensitive position is reached when the object is situated half-way between the light source and the screen (which may be a sensitive emulsion). The scale of reproduction is here 2:1.

This method is hardly suitable for photomicrography, and is only referred to because it is a simple way of observing the phenomenon with small objects and without any special equipment.

It is well-proven that schlieren can be demonstrated by observing the object from one edge between a bright and a dark field. This method can be improved upon by observation of the object through a graticule with parallel transparent and opaque lines, which should be parallel with the schlieren as far as possible. When object and graticule are photographed simultaneously with a sufficiently stopped down lens, the schlieren can be recognized by the shape of the lines of the graticule imaged by the lens. This method has been adapted to photomicrography with the compound microscope in the following manner[32]. The light source is a fine slit in the illumination device of the microscope. This slit is imaged by a condenser and objective in the latter's exit pupil. The graticule is arranged between the exit pupil and the eyepiece. When the lines are prependicular to the slit, the field will show almost uniform brightness, When parallel to the slit, one can observe the shadows of the graticule in the field of view. It is possible therefore to eliminate the graticule design by rotation of one or the other part. When the microscope is focused on a object with details of different refractive indici, the image lines will show distortions corresponding to those described for the macroscopic method. The only difference between the two techniques is that the position of the graticule is in front of the object in one case, and in front of the image in the other. Comparatively simpel apparatus is needed for its realization. The width of the slit should be only a fraction of a millimetre. The graticule should have a 'grating element' of $1/7$ mm, while the recommended ratio of widths of absorbing lines to transparent lines is $1:2$[32].

Fig. 142 shows *Töpler's* arrangement[33], which is the basis of most methods in use. The light source L is an area limited by straight lines, usually on one side, which here lies in the optical axis—but is frequently a slit with adjustable width. This light source is imaged by an optical system SK close to the camera lens towards L'. L' is

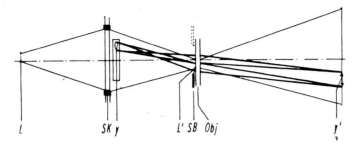

Fig. 142. Schlieren method after Töpler.

then the entrance pupil of the objective for the illuminating rays. The camera lens images the object under observation with schlieren y in the image plane at y'. In the plane of L' is a slideable diaphragm shaped as a knife's edge, the schlieren diaphragm SB, which is exactly parallel with the straight-line limitation of the light source, and which can be inserted in the path of rays at either the lower or the top side, as it appears in the image.

The schlieren deviates the light from its original direction, which strikes it, by refraction and diffraction. It has been assumed in Fig. 142 that the schlieren has a lens-shaped section and gives additional convergence to the light. As long as the light deviated by y passes unhampered through the objective (apart from the usual absorption losses), the same quantity of light will reach the various regions of y' as reaches the surroundings. But if SB intervenes in the radiation area from below, the following changes can be observed in the case of the assumed shape of the schlieren:

1. The image of the top edge of the schlieren is darker than the surrounding, because the light rays coming from the top edge are cut off;

2. The entire image (with the exception of the lower edge of the schlieren) becomes darker, as soon as SB intervenes in L';

3. The image of the lower edge of the schlieren also becomes darker, when the deviated rays which contribute to it, have been vignetted;

4. The entire image becomes dark (with the exception of the image of the lower edge of the schlieren and that of the edges of the test object) as soon as SB exceeds the higher limit of L'.

This process depends on the shape and properties of the schlieren, as well as on the more or less perfectly defined separation of the areas of direct and deviated rays in the plane of the schlieren diaphragm.

It is obvious that the possibility of imaging schlieren which only causes small disturbances—the schlieren sensitivity—increases with the possibility of separation of deviated and direct rays at L'. This sensitivity will increase with increasing focal length of the schlieren head and the distance between object and camera. It also increases with the growing smallness of the light source, provided no unwanted diffraction occurs. The schlieren sensitivity of an instrument used for schlieren photography by means of photomacrography, must therefore be considerably less than that of a specialized apparatus with a schlieren head of often more than 1 metre focal length. It is nevertheless always possible to observe schlieren as produced during electrolysis, solution schlieren, or certain inhomogeneities in solid optical media.

The optical quality of the schlieren head system should be good. It is most important that the light source is imaged in the plane of the schlieren diaphragm aplanatically, and has itself neither schlieren nor inclusions. As far as correction is concerned, photographic lenses or long-focus photo-objectives are quite sufficient. On the other hand, it may be necessary to accept clear spots caused by bubbles as unavoidable.

The further limit of the diameter of the object field (see Fig. 142) is determined by the free diameter of the schlieren head, provided that the illumination device arranged before L produces a sufficiently large illumination aperture (see also the arrangement of AD in Fig. 141).

Various modifications of this method divide the schlieren head into two schlieren lenses separated by a parallel beam of light, in which the object is placed. This results essentially in the same arrangement as shown in Fig. 141. L should be then inserted at AD, and the schlieren diaphragm at AD'. A special additional camera lens is not always necessary. The schlieren diaphragm must, in this case, be arranged in the interior of the camera.

While the modification last mentioned has the advantage of parallel-beam illumination, the first method can be realized more easily with available apparatus, since the use of only two good systems permits the arrangement of the schlieren diaphragm in an easily accessible place. The diaphragm can be simply placed on the camera lens. There is nothing against giving the light source and schlieren diaphragm a circular or annular shape. Such an apparatus is able to image schlieren in all directions in the same way. Fig. 143 was photographed in a parallel-beam path of rays with a knife-edge schlieren diaphragm, while a schlieren head with annular diaphragm was chosen for Fig. 144.

All light sources used in photomicrography are basically suitable. When schlieren in motion is to be photographed, it is advisable that the luminance is sufficiently high. Carbon arc and high-pressure lamps at D.C. are recommended.

Filters do not always have to be used. Their choice depends on the contrast obtained by the apparatus in conjunction with the negative material used. It should be observed that the object does not become fluorescent with arc and high-pressure lamps containing considerable UV radiation. A suitable filter will absorb all blue and UV radiation from the light source, if necessary.

The best type of negative material is one with fairly great contrast, developed in a contrasty working developer.

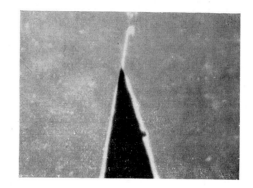

Fig. 143. Electrolytic polishing of a steel point.

Schlieren photograph in telecentric path of rays; $M = 20:1$.

Fig. 144. Schlieren in a fluorite lens. Azimuth-free method with annular diaphragm.

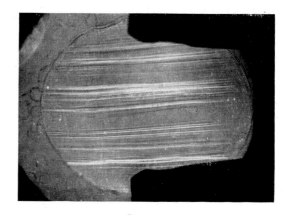

Fig. 145. Same object as in Fig. 144, but photographed by phase contrast.

6.931 Principles. While amplitude objects are usually illuminated with the bright-field technique using a relatively high numerical aperture for good resolution (see 1.634), phase objects are imaged in bright-field illumination with only small apertures of the illumination cone, and even then show relatively little contrast at the edges (see Fig. 146a). The phase contrast method serves to represent phase objects (which are distinguished from their surroundings by only small differences in the paths of rays) with a maximum contrast between bright and dark, but without completely darkening the surroundings, as happens in dark-ground illumination. Small differences in 'optical thickness' which can be present with bacteria and other biological specimens in values of fractions of a wavelength, are converted into differentiated contrasts in intensity. A further advantage of phase-contrast microscopy is that staining is unnecessary. It is therefore eminently suitable for the study of live organisms. This application is also used for the examination of polished metal sections by incident light illumination. The photographs show the various phase shifts and small differences in height of the surface which occur on individual components of the section, producing 'wave' patterns.

Phase contrast causes a reduction in contrast of amplitude specimens. Sometimes, this contrast may disappear completely, or it may be reversed, so that the amplitude contents of the specimen appear brighter than the surround.

According to the definition given in section 6.92, the phase-contrast method is a schlieren method, and is therefore suitable for showing schlieren, as shown in Fig. 145. Contrast is, however, achieved in a different fashion. 'Irregular deflection' in objects of the size of microscope specimens is primarily diffraction. The connection between diffraction and image formation in the microscope has been discussed in section 1.634, and demonstrated in Fig. 34. The diffraction spectrum produced by a phase grid of the same dimensions does not differ from that of an amplitude grid regarding the range and the positions of the peaks, but the maxima of higher order are considerably weaker in light intensity compared with maximum 0. The phase shift of the totality of the deflected light directly behind an individual phase object, and by bright-field illumination at the position of its image, has to be moved by about $\frac{1}{2}\pi$ compared with the light, uninfluenced by the object, when the difference of the optical path lengths through the object and the surrounding field is small. However, the phase shift amounts to π in the case of an amplitude object. Conditioned by the phase shift $\frac{1}{2}\pi$, the wave trains of the direct and deflected components which intersect at the image position cannot interfere to maximum intensification or reduction when a phase object is observed by bright-field illumination (see Fig. 25b).

The phase object produces diffraction images of the light source (the aperture diaphragm of the illumination device), which are normally imaged by the objective at its

---

* Apart from Zernike's phase contrast method, there are others which are based partly on Zernike's method and partly on other principles. The photomicrographic technique is, however, very similar to the one described here, with the exception of a few special colour contrast methods which are rarely used.

Fig. 146.
Ovary of mouse, unstained
paraffin section. Transmitted
light; M = 320:1:

a bright-field, strongly
  stopped down;

b phase contrast, normal;

c phase contrast, strong.

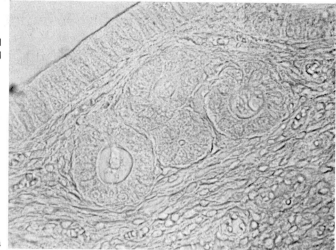

a

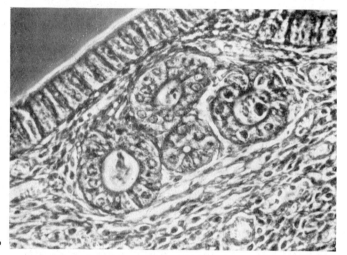

b

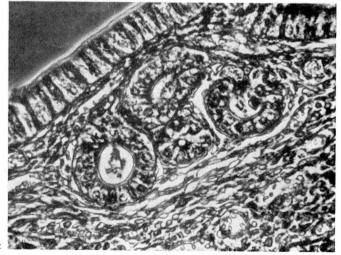

c

rear focal plane. The phase of the diffracted light or of the direct light is so artificially changed that, at the image position of the phase object, the wave trains coming from maximum 0 have a phase difference of approximately π or 0 against those of the maxima of higher order. In the first case we are dealing with 'positive' phase contrast, with the result that the object (which is optically denser than the mounting medium) apears darker than its surrounding. In the second case, by 'negative' phase contrast, the same object would appear brighter than its surrounding. The contrast is also heightened by additional reduction of the intensity of the light which is not influenced by the object. This improves the intensity ratio of the two components.

The phase plate guarantees the change in phase. This is a thin film which is positioned as closely as possible in the image plane of the aperture diaphragm, and which covers this image or its surrounding. It has a different refraction index from the surround. An additional absorption film over the image of the aperture diaphragm assimilates the intensities of the light components which pass through these two areas.

For better efficiency, the commercially available phase plates (and thus the aperture diaphragm of the illuminator) are annular in shape. This ensures a better separation between direct and diffracted light than with a circular section[34].

The phase plate of BEYER's variable phase-contrast method[35] has two concentric 'phase rings' of different diameter and width. The aperture diaphragm of the illuminating device has three ring zones; an iris diaphragm is arranged in their proximity, which obscures the outer zones entirely or partly. When all three zones are used for illumination, the object is seen in bright-field illumination, and the amplitude parts of the specimen can be determined. With two ring zones, we have normal phase contrast combined with a great brightness of field, and excellent resolution and rendering of small phase objects because of the large illumination aperture. According to the large illumination aperture too, larger details will show a halo inside and outside the border to the surrounding, even when the detail is of uniform optical thickness. When using only a single ring zone for illumination (strong phase contrast) these phenomena will not be so obvious, but the object will show other properties inherent in a narrow illumination aperture. Fig. 146 a—c show the same specimen at normal (146 b) and strong (146 c) phase contrast. Comparing these pictures, we find an increase in contrast and a narrower border against the empty field. Fig. 146 a is a comparison photograph taken by bright-field illumination and narrow, central aperture diaphragm, and shows optimum rendering obtainable with this technique.

6.932 Optical equipment. Special objectives which are fitted with phase plates are required for working in transmitted light. They have either a special condenser with a special aperture diaphragm for each objective (arranged in a revolving aperture disc) or a single aperture diaphragm which can be imaged on the phase plate in any desired size by means of a pancratic condenser. A magnifying device is also required to enable the operator to observe the plane of the phase plate when centring the aperture diaphragm on the phase plate. This is done either with an auxiliary microscope which replaces the normal eyepiece, or with a special system which can be swung in the path of rays in the microscope tube. Achromatic or

160

plano-achromatic objectives are used. Special objectives are not necessary for work with phase-contrast incident light, since the phase plates required for the various objectives are arranged at other positions.

6.933 Filters. Optimum requirement can only be met for the wavelength taken as basis for the calculation of the phase plate. It is recommended, therefore, always to use a green filter for obtaining maximum contrast. For particularly contrasty photography, a strong green filter should be used. When a high-pressure mercury lamp is employed, monochromatic filters are excellent. The light emitted by the green mercury line (546·1 nm) is used for illumination.

6.934 Photographic materials. When working with a green filter, fine-grain ortho-chromatic material is always best. A contrasty working developer is recommended since the photographs tend to give an impression of weakness to the unbiased observer in comparison with bright-field negatives when the same negative material is used. This can be explained by the fact that the free background in the negative does not need to be very black, but is usually of a middle tone.

6.935 Technique. Specimens which are particularly suitable for phase-contrast study show sharp and clearly defined contours of the aperture diaphragm (in spite of diffraction) when centring of the diaphragm to the phase plate is carried out. If this is not so, the focused part of the specimen is either too thick (too many super-imposed layers render the image confused), or it has too many amplitude components. Focusing of clear phase specimens can sometimes prove difficult because the object plane is hard to find when the centring of the aperture diaphragm in relation to the phase plate has not yet been carried out. When no amplitude object is to be found in the object plane, the best way is first to focus on the edge of the cover glass, followed by the free object field, and finally superimpose the image of the aperture diaphragm on the phase plate. It is thus much easier to find the object plane. Finally the centring of the diaphragm is again checked. It is obvious that finding the object plane is made much easier when a lower-power objective is used for the first focusing.

6.94 Polarized light photomicrography*

6.941 Principles. The main application of polarized light in microscopy and photomicrography has always been the study of crystalline structures as found in rocks, ores, natural and artificial crystals. But polarized light is also extensively used in metallurgy (determination of the nature of anisotropic crystals) and biology (examination of fine structures). Examination techniques with the polarizing micro-scope are based on the differentiation between objects of different anisotropic properties (direction is dependent on physical properties) by the use of polarized light

---

* Basically also valid for the interference contrast method by polarized light.

(i.e. light, whose light waves vibrate in a single direction or plane). The usual method is to observe the behaviour of the object between two crossed polarizers (arranged with their principal planes at an angle of 90°), which are called polarizer and analyser. The anisotropic effects, often very colourful, are characteristic for the substance under examination, so that it can be identified as a result of this observation. It is also possible to use these anisotropic effects for quantitative analysis. We have no space for a further explanation of polarization effects, which can be found in the literature[36,37]. A simple discussion of the basic phenomena has also been given[38]. It is important for the following considerations to remember that with polarized light photomicrography, the light reaching the photographic material will also be polarized, both linearly and elliptically. We mention this point because ignorance of this has often jeopardized a correct and rational photomicrograph.

6.942 Light sources. Since the coloured anisotropic effects depend on the wavelength of the light, their (colour) character will also be determined by the spectral energy distribution of the illuminant. As these colour effects are used both for photomicrography with only the polarizer (pleochroism, bireflection in plane-polarized light) and between crossed polarizers (interference and dispersion colours) with or without compensators, it is important to determine the principle of the light source.

A daylight-type illuminant was normally used, with its strongest emission at 551 nm, which ensured that the absorption or reflection phenomena occuring in the microscope would correspond to the observation of similar objects in nature. The xenon high-pressure lamp is most useful for polarized light photomicrography, because its extremely high luminance compensates the loss in light transmission by polarization, so that photomicrography with incident polarized light (e.g. ores) requires relatively short exposure times. When maximum luminance is not required, a low-voltage lamp will be used by adapting its colour temperature by means of a day-light filter.

The light source emits an inherently part-polarized light beam. The degree of polarization is, however, very slight. Polarization occurs at the surface of the reflector in the path of the rays, so that the light incident on the polarizer already possesses a preferred direction of vibration. The filament coils of a low-voltage lamp should be perpendicular to the vibration direction of the polarizer, as this ensures the most efficient use of the current. The polarizer, in turn, should be so arranged that its direction of vibration lies on the symmetry plane of the microscope (north-south direction) in the case of transmitted light, and vertical or horizontal (east-west direction of the analyser) depending on the design of the microscope, in the case of incident light. At other adjustments, light losses of up to 20% can occur.

6.943 Filters. A heat filter should always be inserted in the illuminating beam for the protection of the polarizer. Apart from this filter and any other filter required to adjust the colour temperature, a neutral density filter of 2 mm or 4 mm thickness is used, when the required exposure time lies within the camera shake range, and a reduction of the light intensity is therefore required to arrive at a longer exposure. These are the only filters used for colour photography.

For black-and-white photography, the choice of filters will be determined by the colour sensitivity of the photographic material. Thin rock sections show sufficient contrast by transmitted light when an appropriate film or plate is used, thus eliminating the necessity of a contrast filter. The same is valid for ore specimens between crossed polarizers. On the other hand, ore specimens in plane-polarized light (polarizer only) by reflected light will show mainly different tones of grey and very little colouring. The individual grain contours are frequently blurred in unfiltered light, so that the use of a filter is recommended when negative material of steep gradation does not give the desired result. Monochromatic filters are unnecessary. The filter set required for this type of work would consist of blue, green, yellow-green and orange filters. Table 16 gives a few applied examples for the differentation of ore minerals with these filters.

A filter can often be discarded when photographing biological specimens. When a suitable negative material is selected, sufficient contrast will be obtained in the first-order greys of the anisotropic objects. If more contrast is required, a fixed or adjustable mica compensator can be used ($^1/_4$—$^1/_{16}$ wave plate). For higher-order polarization colours, contrast filters are selected in the normal way, if required (see **3.611**).

6.944 Optical equipment. The objective used for polarized light photomicrography must be free from all traces of strain birefringence, since any strain will considerably falsify the character of anisotropic effects (strain birefringence). Achromatic objectives which are as free as possible from strain are recommended, since higher colour correction can only be achieved by using several fluorite lenses, which often have inherent anomalous birefringence. Apochromatic and fluorite objectives should therefore not be used. It is a fallacy to assume that the strict requirements of freedom from strain would only apply to quantitative microscopy. Although it is possible to eliminate chromatic aberration (which would be unwanted in colour photography) by means of an apochromatic objective, anomalous birefringence above mentioned can falsify the anisotropic effects beyond recognition, especially in the case of ore specimens. The best photographic quality is obtained with a strain-free plano-achromat, which yields outstandingly flat image planes in conjunction with compensating or plano-compensating eyepieces (Fig. 147). Achromatic objectives with Huygens eyepieces can be used if there is no objection to residual curvature of field.

6.945 Photomicrographic equipment. Any photomicrographic apparatus is basically suitable for work with polarized light, if the following instructions are observed. The inherent light loss favours the miniature size, particularly by reflected polarized light. The medium-sized camera can be used in a few cases, although exposure times of 10 to 30 minutes are no exception here, and are always longer than one minute. The miniature camera is attached to the microscope as described in section 4. There is no change in the use of a micro adapter, special stand or fixed meacra lens. The only exception is the attachment camera if it is fitted with a beamsplitter. We have already mentioned that the image-forming light beam is plane-polarized. The beamsplitter acts as partial polarizer and can exert such an influence

Fig. 147. Stibioluzonite from Candalosa mine, Peru. Incident light—bright-field, crossed polarizers
$M = 20:1$ original; enlarged to $M = 80:1$. Agfacolor UT 16.

on the image-forming light that anisotropic effects are entirely distorted. A further point is that the image observed in the focusing system does not correspond to that recorded on the film (because of the different polarizing planes) and that both are different from the true relationships in the specimen[40]. Fig. 148 shows a typical example. A thin granite section with pleochroitic minerals was photographed by a camera attachment with beamsplitter. Only the polarizer was inserted in the path of rays. The focusing eyepiece showed an image (Fig. 148a) similar to that between crossed polarizers (with a small degree of polarization). The actual relationship is represented in Fig. 148b. The image recorded on the photographic material would give the effect of parallel polarizers. But the small degree of polarization is usually sufficient to render this effect almost imperceptible.

The above distortions can be almost fully suppressed by following this technique:

1. Adjust the analyser so that its direction of vibration is perpendicular to the optical axis of the focusing system;

164

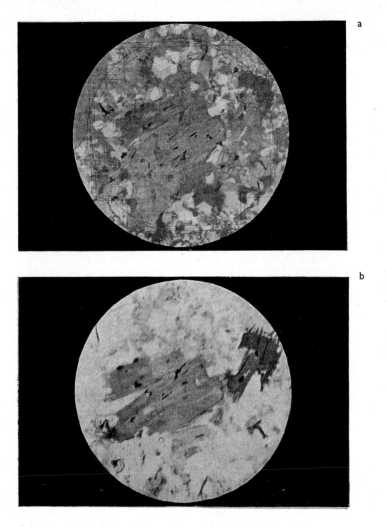

Fig. 148. Granite from Henneberg. Transmitted light—bright field, 1 polarizer $M = 32:1$ original enlarged to $M = 100:1$. Agfacolor UT 16.

2. When pleochroitic objects are photographed, do not use the polarizer (before the specimen) but the analyser (after the specimen). The latter should be adjusted in relation to the focusing system as in 1.

It is absolutely necessary to follow this technique if the photograph should portray the actual relationship. Rotation of the attachment camera over the analyser is out of the question. One could be tempted to apply this when using a miniature camera, for example, for the inclusion of a group of crystals which is situated in a defined position with regard to the vibration planes of the polarizers. The above instructions should be ignored when an attachment camera with swing-out, total-reflection prism is used.

6.946 Photographic materials and processing. In colour photography, the type of colour film used is determined by the illuminant. This means that only daylight colour film can be used. There are no general rules for black-and-white materials, since all that is required is the adaptation of the material to the method in the best possible way. Panchromatic materials are only essential when it is required to differentiate between pink or red, and grey or black minerals. Orthochromatic materials give the best results, since the achromatic objectives mainly used are insufficiently corrected with red light for spherical aberration. It is often necessary to encompass very large contrasts in the photomicrography of rocks and ores in crossed-polarized light (Fig. 149). A low-contrast emulsion (which will almost always be panchromatic) is recommended, which can be developed in a good 'compensating' developer. On the other hand, with the frequently weak anisotropic structures in biological speci-

Fig. 149. Thin stone section in cross-polarized light Norite from Bamle, Norway. Transmitted light—bright field; $M = 40:1$ original, enlarged to $M = 160:1$. Agfacolor UT 16.

166

mens low-contrast emulsions do not render the small differences in grey tonalities strongly enough[41]. Emulsions of a much steeper gradation are necessary here, which will yield brilliant negatives after development in a universal or compensating developer. The negatives are then printed on normal or soft paper, and give an excellent reproduction of the visual impression, apart from being eminently suitable for block-making. The same considerations apply to ores photographed in reflected polarized light, for which case hard-working photographic materials are recommended. This should not be generalized. Survey photographs of biological specimens with crossed polarizers may require the use of extremely soft-working materials and developer (Fig. 150). Ore sections with an appreciable amount of gangue minerals also require a soft-working negative material. Isopan F has given excellent results for exposure through a green filter, developed in Rodinal 1:20—1:40 for plates, and in Atomal for films. It was possible to show the individual gangue types in different shades of grey.

Axial image photography requires very hard-working materials, which can be developed either in a commercial metol-hydroquinone developer (diluted 1:4; 4 to 5 minutes at 20 °C) or, better still, in a developer of the following formula:

Metol	0.8 g	Potassium carbonate (anh)	50	g
Sodium sulphite (anhydrous)	40 g	Potassium bromide	10	g
Hydroquinone	8 g	Water to make	1,000	cc

Development time is 6—7 minutes at 20 °C.

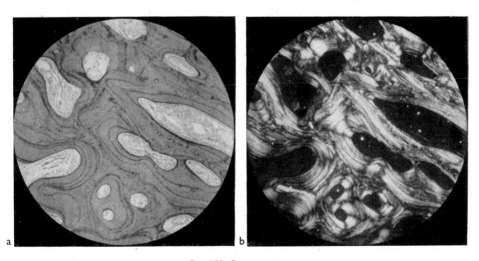

Fig. 150. Bone section

Transmitted light—bright field; $M = 25:1$: *a* without polarizers; *b* with crossed polarizers

Because of the very strong brightness range in the object, fine detail could only be rendered satisfactorily by using a soft-working panchromatic plate (Isopan F) and developing in Rodinal 1:80 (6 minutes).

167

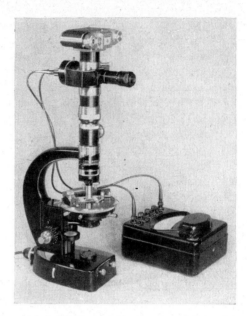

Fig. 151. Microscope for polarized light with photomicrographic attachment and exposure meter.

6.947 Special techniques

6.9471 General considerations. A set-up analyser is often unsuitable for photography. It is difficult to rotate the analyser or to alter its position in relation to the attachment camera, because in practice the analyser is often no longer accessible in the complete arrangement. Another annoying point is that at the small diameter of the exit pupil, any dust in, or on, the analyser will become visible in the photograph. For this reason, tube analysers are to be preferred. Special polarizing microscopes are always fitted with these (Fig. 151).

When making colour photographs of birefringent objects with objectives of high numerical aperture, the aperture diaphragm should be closed down until the first traces of diffraction fringes appear. Otherwise, the colour rendering for parallel light will be impaired, and the image may give the impression of a higher degree of birefringence than was actually present in the original. When photographing ores between crossed polarizers, the image of the aperture diaphragm, which will be considerably stopped down, should be as close as possible to the edge of the prism (observation with Bertrand lens), so that the angle of incidence approaches as near as possible to 0°. The N.A. of the illumination should not exceed a value of between 0·15 and 0·17.

6.9472 Axis-image photomicrography. Photographic reproduction of axis images produced in the rear focal plane of the objective can be difficult, especially when the axis image of a crystal should be produced from a thin section. Good results can

168

only be expected from a most careful adjustment of the illumination. The even illumination of the image field depends on the exact centring of the field diaphragm. This centring is best carried out immediately after the image of the exit pupil of the objective, after having previously adjusted the illumination in the normal way by direct observation of the object. With a narrow aperture of the field diaphragm, crescent-shaped shadows (most frequently blue in colour) will appear in the exit pupil. Appropriate tilting of the illumination mirror can change this crescent shape to a circular one concentric with the demarcation of the image field. The thus centred field diaphragm can then be opened up.

Since condensers used in polarizing microscopes can have optical aberrations (particularly spherical aberration), it is necessary to deviate from Köhler's illumination method, to open the field diaphragm further than usual and to re-focus the condenser if necessary, until the entire exit pupil of the objective is evenly illuminated. Otherwise, the image will be vignetted by the field diaphragm, and surrounded by unsharp fringes of a reddish colour. Since immersion objectives are usual for the making of axial images, it should be remembered to arrange an immersion layer (preferably oil) between condenser and object carrying slide. The axis image is produced in a plane which represents the image of the light source in Köhler's illumination. The most frequently used low-voltage filament lamp is often of a relatively uneven structure, resulting in a lack of uniformity in the illumination. Good illumination can be obtained by focusing the light source against the bull's-eye condenser so that its image is shifted in relation to the object plane. A ground-glass screen should always be inserted in the path of rays.

For photomicrography by indirect observation, the miniature camera is most frequently used as an attachment camera or with a micro adapter.

Axial photography with standard microscopes and polarizing accessories needs an additional attachment because of the lack of an Amici-Bertrand lens. The auxiliary microscope used for phase-contrast work can be used in case of emergency. This will necessitate putting on and taking off the camera several times for optimum focusing, because the auxiliary microscope can no longer be reached when the camera is in position. The same occurs with the attachment camera. It is easier to carry out visual focusing (with the relaxed eye or pocket telescope) and to place an 8-diopter spectacle lens over the auxiliary microscope prior to photography. The axis image will then be sharp on the film plane.

By using modern microscopes with rapidly interchangeable tubes, it is possible to make an auxiliary microscope by means of a draw tube, in whose lower end a Zeiss "M" lens of 45 mm focal length is screwed. The choice of the projection system depends on the size of the photographic material. A universal eyepiece $\times 7$, or a projection eyepiece 4:1 is very suitable in conjunction with the reproduction lens mentioned above. The lens is then stopped down to an aperture of 2—3 mm. It is thus possible to carry out focusing with the camera in position by shifting the draw tube. When the optimum position has been established, auxiliary microscope and camera are locked into position by sliding a locking ring over the tube. As the whole arrangement is somewhat unstable, the scale of reproduction of the axis image should be so chosen that the entire film frame area is not completely used. Otherwise part of the image may be cut off at the slightest upset of the adjustment.

The objective for reproduction of an axis image on colour miniature film should be selected on the ground of its N.A., so that it does not take up more than three orders in the image field. Although an objective with higher N.A. would encompass a larger image angle and hence also a greater number of interference orders, colour film is unable to give a true rendering of isochromats of the higher order. Colour rendering is poor even in the third, and often in the second order. Green[II], blue[III] and yellow[III] are often rendered as white strips. The causes can be traced back to the following phenomena: With the daylight-type illumination used in polarized light photomicrography, the already intense green and blue hues are greater intensified because of the strong green and blue contents of the light. On the other hand, not all colours tend to become whitish to the same extent by an increase in exposure; yellow hues become whitish sooner than others[42]. This means that there will be no yellow of the higher order (at increased absolute intensity of the isochromates), if exposure has been calculated for the intensity of the isochromates of the first

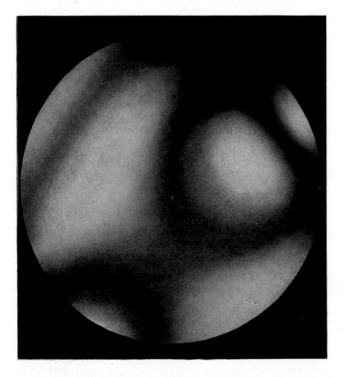

Fig. 152. Axial photograph of a forsterite, near-parallel (100) from a thin section of olivine basalt. Crossed polarizers, crystal in diagonal position. Agfacolor UT 16. The exit pupil of the plano-objective used 100/1.25 H I was imaged by using an auxiliary microscope consisting of an objective $M = 1:4.5$, $f = 45$ mm and a projection eyepiece 4:1.

170

order. Only the first of these sources of error can be partly eliminated by choosing a light source with higher red content (artificial light) and hence an artificial light colour film. The second source of error would hereby be intensified. Moreover, practice has shown that colour rendering is better with daylight colour film than with the other type, so that daylight film should also be preferred for axis image photography. Optimum colour rendering is obtained by the shortest possible exposure time (Fig. 152).

6.95   Infra-red photomicrography

6.951   Principles

6.9511 Applications. Microscopic studies in the infra-red region of the spectrum used to be solely by photomicrography. Means for direct observation of the infra-red image were introduced later. The development of infra-red photomicrography commenced relatively recently (*Köhler*, 1912) as the desire for maximum resolution in photomicrography brought with it the preference for UV radiation. It is true that the losses in resolving power in infra-red work are quite considerable. For example, an objective with a N.A. of about 0·80 will resolve a structure with a grid distance of 0·35 micron by illumination by green light (550 nm); see equation (22c). The same structure in UV (257 nm) will already be resolved by objectives with a numerical aperture of 0·40, but in infra-red (920 nm) only by immersion objectives with a numerical aperture of 1·30. Because of these natural relations, infra-red radiation also produces a considerable displacement of the range of useful scale of reproduction (Fig. 153), which should be taken into account when selecting the optical equipment.

The importance of infra-red photomicrography lies in the fact that the representation of the absorption relations in the object yields essential information on its morphological structure. The applications have become increasingly wider and more varied, because many substances, which are partly or wholly opaque in the visible spectrum, show less absorption in the short-wave infra-red region and can therefore be studied by transmitted light. Its advantage is that the frequently necessary bleaching of specimens can be omitted. Apart from less work being involved, infra-red photography also guarantees a greater objectivity, since the specimen is not subjected to chemical reactions which can sometimes lead to morphological changes. For instance, in the microscopy of textiles, dyed samples can be studied directly, because the dyes used in the textile industry nearly all transmit infra-red light[43]. Insect chitin is almost opaque in visible light, but transmits short-wave infra-red radiation (800 nm). Fossile chitin (graptolites[44]) shows the same absorption properties, so that palaeontological research makes frequent use of infra-red photomicrography.

Other organic applications are the infra-red photomicrography of healthy and diseased human skin (capillary photography), diagnosis of cataracts and other diseases of retina and iris, as well as comparative anatomy of human hair[45]. In the

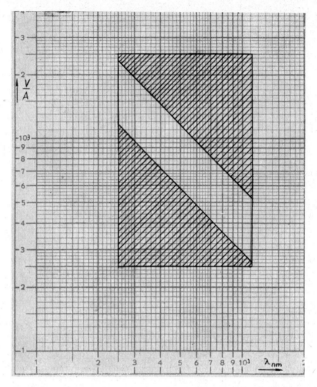

Fig. 153. Useful magnification V/A against wavelength of microscope lamp illumination. The scale of reproduction of the final print should be such that the ratio M/A lies within the area encompassed by the two straight lines.

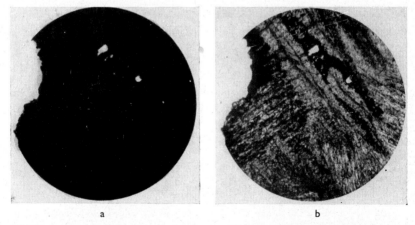

a   b

Fig. 154. Brown iron ore from Auerbach, Bavaria. Transmitted light—bright field; $M = 50:1$.
$a \; \lambda = 550$ nm Agfa Micro plate, $b \; \lambda = 850$ nm Agfa Infra-red plate 850 hard.

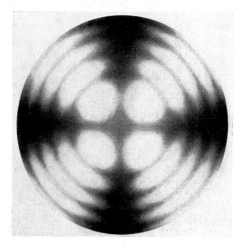

Fig. 155. Axial photograph of a molybdenite cleavage plate, parallel (0001). $\lambda = 850$ nm; interference filter.

case of skin diseases, such as bovine and Norwegian scabies, the underlying tissue can be observed through the external manifestation.

The absorption decrease, starting in the short-wave infra-red region, of a number of opaque minerals (sulphides and oxides mainly of elements of the IVth and Vth period, as well as a few sulpho-salts) allows the study of morphological and crystal-optical relations by transmitted light (Fig. 154). This is doubly advantageous from a polarization-optical point of view, since on one hand transmitted light techniques are not so sensitive to adjustment, and on the other hand quantitative methods are less complicated than those for reflected light techniques. Axis images (Fig. 155) also, can be obtained and interpreted more easily by transmitted infra-red radiation than by incident light[46].

6.9512 Detectors. One can distinguish between direct and indirect infra-red techniques according to the type of detector being used. The direct method employs infra-red plates or films in the normal photomicrographical apparatus. This is the simplest and cheapest infra-red technique, but has the drawback of the operator not being able to see what he is doing since the eye cannot perceive the infra-red image. This necessitates a number of test exposures before a successful result can be obtained. One of the oldest infra-red techniques has the same disadvantage. It is an indirect method, based on the Herschel effect, where infra-red radiation has the property of bleaching the latent image formed in fast photographic plates[47] or silver bromide papers[48]. The photographic material first receives a diffuse pre-exposure, immersed in certain sensitizers (most frequently weak solutions of green dyes), washed and dried. By exposure in the photomicrographical apparatus, a positive, laterally reversed image of the specimen, illuminated by infra-red light, will be obtained. Infra-red radiation has generally a stronger effect on poorly exposed spots than on strongly exposed

173

ones. When therefore the diffuse pre-exposure is not too strong, the necessary infrared exposure can be shortened—at the expense of a loss of contrast.

Although this method has four serious disadvantages ('blind' operation, a positive which cannot be printed, laterally reversed presentation, and long exposures) it is however sometimes used, particularly when infra-red work is undertaken only sporadically and where no infra-red materials are available. The second indirect method has greater importance but requires more instrumentation. It uses an electronic image converter[49], where the infra-red image is formed on the infra-red sensitive converter cathode and imaged on a fluorescent screen by means of an electrostatic or electromagnetic lens system. The high voltage (up to 25 kV) which is applied causes the photo-electrons emerging from the photo-cathode to impinge on the fluorescent screen with such a force, that the latter is excited to fluoresce.

The screen then shows an image whose brightness distribution corresponds to the infra-red transmission of the object. The great advantage of the image converter is that the screen image can be visually examined allowing a determination of image framing and scale of reproduction without having to make test exposures. Another advantage is that by varying the kilovoltage, additional energy can be brought in the path of rays, resulting in a much brighter image than obtained with the direct method. The long-wave sensitivity limit is reached at about 1,100 nm; this applies to all methods.

6.952 Radiators. The high infra-red content of the light of a low-voltage filament lamp would seem to render this type of light source suitable for infra-red photomicrography. But exposures are six times shorter with the universally applicable carbon arc lamp. The best radiator is, however, the xenon high-pressure lamp which needs much less attention than the arc lamp. By optimum utilization, the infra-red content of the carbon arc lamp is three-fifths of that of the XBO 100 xenon high-pressure lamp. The zirconium arc lamp can also be used, but does not reach the intensity of the xenon high-pressure lamp.

6.953 Filters. For survey work there are filters with transmission in the total infra-red region between 700 and 1,000 nm with the preferred wavelength 850 nm, e.g. the combination of Schottglass BG 3 and RG 5, both 2 mm thick. Fig. 156 shows the transmission curves of a few infra-red filters. Interference filters are used for more detailed observations. The line spectrum of the high-pressure lamp should correspond to the colour sensitivity of the photographic material. When using the XBO 100 high-pressure lamp, a set of interference filters with transmission maxima at lengths of 760, 820, 850, 900, 1,020 and 1,080 nm is recommended. The maximum degree of transmission of this type of filter is averaged at 0·30; the half-value width lies at 10—15 nm.

Any heat-absorbing filters must be removed, as they absorb infra-red radiation.

6.954 Optical equipment. Achromatic objectives are not recommended because of their lack of spherical correction for red and infra-red rays. Apochromatic objec-

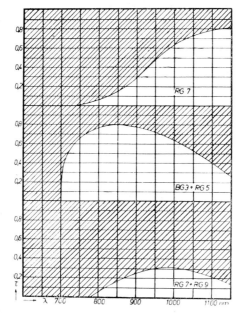

Fig. 156. Transmission curves of some infra-red filters.

tives are better, because the difference in focus between red and infra-red rays can be ignored with them in practice. Mirror objectives, however, are best in this respect, as they do not have any focus difference. Unfortunately, mirror objectives are not yet available for all the required scales of reproduction.

Infra-red absorption by lens-objectives can also be ignored.

The choice of projection system is comparatively free. As infra-red photography mainly works with monochromatic light, correction for chromatic difference of magnification does not have much importance. Apochromatic or mirror objectives can therefore be used with eyepieces or projection eyepieces with or without compensation.

6.955 Photomicrographic equipment. Plate cameras are best used with the direct method of infra-red photography, because infra-red materials are mostly available in plate form. For short exposures, an attachment camera for 6.5×9 cm plates is recommended. Its greater brightness in the focal plane guarantees more exact focusing than with larger size cameras. Use of the miniature camera is limited by the available infra-red film to the short-wave infra-red range up to about 780 nm.

With an image converter, the screen can be photographed with a miniature camera with extension tube or bellows. The scale of reproduction given by the camera lens should be about $1.4:1$. This compensates the magnification factor of the image converter (mostly 0·7), while the 24×36 mm frame will be fully illuminated (image size on the screen is approximately 35 mm in diameter).

6.956 Photographic materials. A range of infra-red plates of different peak sensitivities is available for direct infra-red photomicrography.

Kodak infra-red extra-rapid plates are available in the sizes $2\frac{1}{2} \times 3\frac{1}{2}$ in., $3\frac{1}{4} \times 4\frac{1}{4}$ in., 4×5 in., $4\frac{3}{4} \times 6\frac{1}{2}$ in., 6.5×9 cm and 9×12 cm. Kodak Class R infra-red plates have a peak sensitivity from 740—850 nm and are recommended when high contrast is required. Class M plates are useful from about 860—1,000 nm (maximum at 930 nm). Class Q is sensitive from 720—1,100 nm with its maximum at 980 nm. Class Z extends further into the infra-red; it has a maximum at 1,085 nm and has been used successfully beyond 1,200 nm. Kodak Class M, Q and Z must be hypersensitized before use by soaking for 1 minute in a solution of 4 cc of 0·880 sp.gr. ammonia in 100 cc of water, at about 16°C, and drying as rapidly as possible in a current of air.

For photomicrography, a low-contrast developer is recommended. In the case of Kodak infra-red extra-rapid plates, development for 4 minutes at 20°C in Kodak DK-50 developer (dilution 1 : 1) yields a gamma of 0·8.

For the photography of the image converter screen, a panchromatic film, such as Kodak R. 55 recording film can be used to record the characteristic green colour of the fluorescent screen. But the special Kodak "Flurodak" panchromatic film has a very much greater speed (and hence needs shorter exposures), combined with a very fine grain, exceptional for such a high speed. Development is usually in Kodak D. 19 C developer (6—8 minutes at 20°C).

The difference between exposure times for the direct and indirect methods is very great. With the direct method, exposures of 30 minutes are no exception, while the fluorescent screen may only need 1—4 seconds.

6.957 Special technique. The image quality obtained with the image converter is generally inferior to that of direct-method negatives. This is caused by distortion introduced by the image converter. If the highest possible quality is required, the image converter will be used for visual observation, while photography is carried out by switching to the regular photomicrographic apparatus with infra-red plates.

Superficially, one may think that focusing will be very difficult since the radiation is not visible. But using the image converter makes it extremely easy as the sharpness can be accurately adjusted. Focusing with the direct method with mirror objectives is just as easy, since no focus difference occurs. When an apochromatic objective is used, it is recommended to focus on a clear-glass screen by dark-red light (filter). The focus difference between this light and the infra-red radiation is so small that it does not influence sharpness. If an achromatic objective should be used in an emergency, the focus difference can be corrected as follows: First focus by green light (green filter) and note the position of the fine-focusing knob. Do the same with a red filter. Subtract the second reading from the first, double this value and rotate the knob by this amount in the direction of the optimum sharpness for red. This will give correct focus for a wavelength of approximately 820 nm. This technique gives excellent results, provided the fine focusing mechanism is in good order.

Exposure time is preferably determined with test exposures. A photoelectric meter is used with the fluorescent screen method. For direct photography, infra-red sensitive photo-cells (with caesium oxide cathode) are required, which need an auxiliary power supply, and also great care in their use. The instrument should have a sensitivity of at least 10^{-9} amp for graduation.

6.96   Ultra-violet photomicrography

6.961 Applications and detectors. Compared with green light illumination, ultra-violet radiation can double the resolution of some objects. The limit of resolution for wavelength 257 nm is 0·11 μ, for wavelength 550 nm it is 0.20 μ. But the gain in resolution is small compared with the increase in size of apparatus. The widest field of application of UV photomicrography—similar to infra-red work—lies in the representation of object structure and object component on the basis of differential absorption capacity. The absorption differences of the individual object components show an inherently sufficient contrast in the photomicrograph, so that this technique can be used for the photography of live objects, without any special techniques such as staining, and phase-contrast, provided the UV radiation itself does not have a harmful effect on the object. The varying content of albumens in biological specimens determines absorption differences. UV photomicrographs can therefore be used for the determination of both qualitative and quantitative distribution of such substances in tissues.

We also distinguish between direct and indirect methods in UV photomicrography as with infra-red photomicrography. The direct method works with specially sensitized plates; the indirect, with image converters. The only difference between a UV image converter[52] and an infra-red one is the different spectral sensitivity of the photo-cathode.

6.962 Radiators. A thermo-radiator cannot be used for UV photomicrography. Only the carbon arc lamp can be employed in the long-wave UV region. Mercury high-pressure lamps give considerably better results, because their discreet UV spectrum allows a better monochromosity of the illuminating radiation. A hydrogen lamp can also be used, but requires a monochromator for each defined monochromosity. The most intensive UV radiators use H.T. sparks between cadmium, magnesium or zinc electrodes. This requires a considerable amount of apparatus[53], but has the advantage of the most exact operational possibility. Spark radiators are generally reserved for use in special UV equipment.

6.963 Filters. The means for isolating particular spectral regions depends on the type of work. The use of a monochromator is unavoidable for the highest requirements, when a precisely determined wavelength is allocated for the UV absorption of object components. As its light intensity tends to be low, it should be used in conjunction with a spark radiator. Filters are suitable for survey work, particularly in combination with mercury lamps. The monochromaticity is sufficient despite a relatively wide transmission band. An interference filter of up to 330 nm can be employed. *Meyer-Arendt*[15] has described the following filter set, consisting of glass, liquid or gaseous filters.

$\lambda = 366.3 + 365$ nm	UG 2/2 + BG 12/2
$\lambda = 312.6 + 313.2$ nm	1) UG 5/3 + WG 5/4
	2) nickel sulphate heptahydrate solution (68 g in 100 cc of water) in cells of 60 mm.
$\lambda = 280\cdot4$ nm	1) UG 5/3
	2) nickel sulphate heptahydrate solution (51 g in 100 cc of water) in cells of 30 mm.
	3) picrinic acid (63 milligram in 1,000 cc of water) in cells of 10 mm.
$\lambda = 253\cdot7$ nm	See 3·624.

6.964 Optical equipment. Achromats and apochromats cannot be used because of their poor correction for UV rays and strong UV absorption. Normal photomicrographic equipment with apochromats can be used for wavelengths down to 300—320 nm. It is, however, better to use mirror objectives, because of their lack of focus difference. However, they are not made in all the required numerical apertures, which reduces UV photomicrography to monochromats. It is obvious that other optical elements in the system (bull's-eye condensers, filter cells, plane-parallel glasses, condensers, object slides, cover glasses, immersion media and eyepieces) should all transmit UV radiation. All the elements should therefore be made of molten quartz. Rock crystal is less suitable because of birefringence. Quartz bull's-eye condensers, condensers and eyepieces are available commercially. The illumination mirror must be replaced by a quartz total reflecting prism, or by a surface-aluminized mirror of molten quartz. The reflection degree of this mirror is almost constant at 0·91 between 253 and 405 nm[54]. Glycerine is used as immersion medium.

6.965 Photomicrographic equipment. Medium or large size bellows cameras are used for direct photography. Attachment cameras are not recommended because image focusing with the focusing system is not possible, and the most frequently built-in beam splitter would absorb all radiation shorter than 320 nm. With regard to the image converter, the equipment described in section 6.955 is used.

6.966 Photographic materials. The silver bromide of non-colour-sensitive (and also of colour-sensitive) plates has an inherent sensitivity for radiation between 200 and 400 nm. This makes their use for UV photomicrography possible. The sensitivity can be increased by coating the emulsion with a thin layer of fluorescent paraffin oil prior to exposure. This transforms the short waves into longer ones which can penetrate more readily the gelatin of the emulsion. *Ross*[68] recommends a mixture of paraffin oil with acetone, benzol or ether. This mixture is carefully applied on the surface with cotton-wool, and removed before development by means of a suitable solvent, so that the developer has free access to the emulsion. Our own experiments have shown that this method of increasing the sensitivity is unsuccessful in the area beyond $\lambda = 300$ nm. The method is cumbersome, and the risk of damaging the plate does not bear any relation to the gain in sensitivity. Since all such plates show decreasing contrast by a decreasing wavelength of the incident light (Fig. 157[55]), it is recommended to compensate this by using high-contrast plates, such as the Kodak Uniform Gamma plate for UV spectrography. Specially sensitized plates are to be preferred, e.g. Kodak Scientific Plate III-0, UV sensitized, particularly when working in the region shorter than 300 nm. These plates have the advantage of almost constant contrast throughout the entire wave-band used.

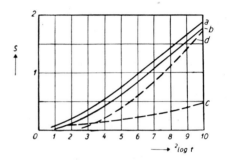

Fig. 157. Characteristic curves of some UV negative materials.

a non-colour-sensitive plate ⎫
                                 ⎬ 300 nm
b UV plate   ⎭

c non-colour-sensitive plate ⎫
                                 ⎬ 230 nm
d UV plate   ⎭

6.967 Special technique. Focusing demands special mention. It cannot cause difficulties when a mirror objective or an image converter is used. The greatest difficulty in focusing is overcome when a monochromat is used. Focusing by visible light and subsequent correction are impossible, because no definite focus can be determined by visible light, owing to of the correction position of the monochromat. The multi-exposure device can be used with different positions of the fine-focusing knob, but this empirical method is cumbersome and time consuming. Replacing the ground-glass screen by a fluorescent screen is also of doubtful value because of difficulty in focusing by the low intensity of the fluorescence.

The best method is to use Köhler's UV viewfinder, which produces an image in the plate plane reduced to one tenth of its size and therefore 100 times brighter. This image can be focused with a magnifier. An adjustment device compensates the aberration when different wavelengths are used.

The scale of reproduction of monochromats with quartz eyepieces is calculated in a different way to that which has been described so far. Monochromats are characterized by their focal length f_{obj}, and quartz eyepieces by their magnification $Mg_{ocular}\left(Mg_{ocular} = \dfrac{\Delta}{f_{obj}}\right)$. The scale of reproduction for a camera extension k is then—

$$M = \frac{k \cdot Mg_{ocular}}{f_{obj}.} \tag{36}$$

This equation also allows the determination of the camera extension for a given objective-eyepiece combination. Because of the correction of quartz eyepieces and the tuning of the UV viewfinder, it is recommended to use a camera extension of 30 cm.

6.97 · Fluorescence photomicrography

6.971 Principles. Fluorescence can be defined as the property of the atoms or molecules of certain substances to absorb radiation of a particular wavelength and to re-emit it as light of longer wavelength. Microscopic objects also show this property, but fluorescence photomicrography became increasingly important only through the progress made in fluorescent materials. With these methods, the specimens are stained with fluorescent substances (coriphosphine, auramine, acridine orange, etc.). Depending on the concentration and degree of adhesion to the tissue (molecular or electrostatic), a range of different hues can be obtained from a single dye. The method is far superior to other methods of staining in this respect[56].

Absorption of the excitation radiation and emission of the fluorescent light nearly always take place in continuous bands. Because of the position of the absorption bands of the dyes used for fluorescent microscopy, both long-wave UV rays and short-wave blue are suitable for the excitation. The fluorescence bands then lie in the green, yellow or red regions.

It is immaterial how the excitation radiation reaches its target. For the sake of greater light utilization working with straight bright-field illumination is best. Oblique bright-field illumination (possibly to obtain greater resolution power) is of no great advantage, since in fluorescence microscopy the image formation laws for luminous specimens are valid, and the type of path of rays through the illumination space has no influence on the resolution. It is, however, important to use Köhler illumination. The size of the illuminated object field should be limited exactly to the part of the specimen to be photographed (by means of the field diaphragm). This is very important since many specimens will fade during intensive UV radiation, which will gradually destroy the illuminated area of the specimen (Fig. 158).

When reflected light is used, plane-parallel glass is best for fluorescence photomicrography. The usual built-in prism is made of glass of a high refractive index (favouring polarization) and this prism will absorb a considerable part of the excitation radiation. Quartz bull's-eye condensers, condensers and object slides are unnecessary since the glass components of the normal photomicrographic apparatus do not greatly absorb long-wave UV radiation. But specialized fluorescence photomicrography apparatus has bull's-eye condensers made of UV-transmitting glass. Care should be taken that none of the optical components of the apparatus has autofluorescence. This is particularly so in the case of immersion-oil for high-power objectives. Normal immersion oil cannot be used, but is replaced by special fluorescence-free immersion oil.

180

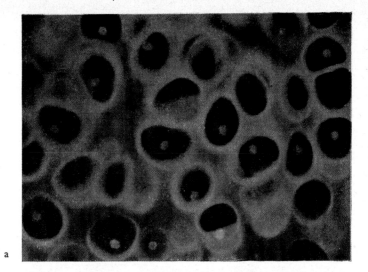

a

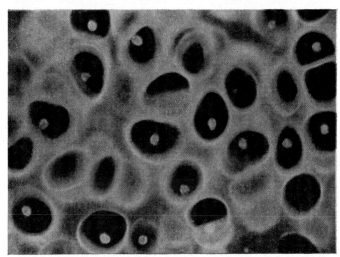

b

Fig. 158.
Fading of a specimen stained
with acridine orange.

Blue-light fluorescence with
BG 3.
Bronchus of guinea pig.
$M = 100:1$ original
enlarged to $M = 250:1$.
Agfacolor UT 16:

a after an irradiation
of 3 minutes;

b after an irradiation
of 5 minutes;

c after an irradiation
of 9 minutes.

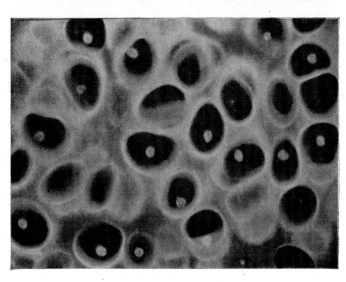

c

6.972 Light sources. The position of the absorption bands of fluorescent substances demands that the light source must emit high proportions of UV and blue. The carbon arc lamp is the simplest strong light source which fulfils this requirement. Cored carbons should not be used as they tend to splutter and will coat the bull's-eye condenser in a short time. Mercury lamps are even better than carbon arc lamps. The most intensive excitation is obtained with the mercury high pressure lamps (HBO 50, HBO 200), whose arc dimensions most favourably correspond with the requirements in the microscope. In exceptional cases, such as the diagnosis of tuberculosis by fluorescence photomicrography, the short-wave content of the radiation emitted by low-voltage lamps will suffice. A 100 watt projection lamp will be excellent for this purpose.

6.973 Filters. Filters are also extremely important in fluorescence photomicrography. They can be separated into two groups—transmission filters and absorbing filters. Transmission filters must match the radiation of the light source. The liquid filters used earlier are now replaced by glass filters. The UV transmitting filter is inserted anywhere below the object, while the UV absorbing filter is placed on the eyepiece. Both filters are available in different degrees of absorption and transmission so that any desired combination can be effected. The Kodak "Wratten" 2 B filter (very faint yellow) is the most frequently used absorbing filter.

Various absorbing filters are sometimes arranged on a rotating disc in the microscope tube. With reflected light illumination it is possible to make the plane-parallel glass also serve as an absorbing filter[57].

6.974 Optical equipment. To obtain optimum brightness of the fluorescence image, both condenser and objective should have the highest possible numerical aperture for the corresponding focal length. Thus use of aplanatic or achromatic-aplanatic condensers with a numerical aperture of 1.4 in combination with an apochromatic objective, is to be preferred. So far no auto-fluorescence of the fluorite lenses used in apochromatic objectives has been observed. The high-aperture Zeiss Apochromat 40/0·95, corrected for a cover glass thickness of zero is particulary suitable for fluorescence photomicrography of smear specimens. Apochromatic objectives are also advantageous with reflected light illumination because of their brighter images. Only compensating eyepieces should be used in this case.

6.975 Photomicrographic equipment. Despite judicious selection of filters and optical equipment, the fluorescent image shows an extremely low level of brightness, requiring very long exposures. It is therefore not advisable to further increase these exposures by using large size cameras (see 3.4). In principle, the attachment with the miniature camera is best. For single exposures, a plate camera with minimum extension can be used. Single-lens reflex cameras with micro adapter are only of advantage when they can show the image on a clear-glass screen for focusing. Ground-glass focusing does not guarantee sufficient accuracy in focusing and cannot be recommended.

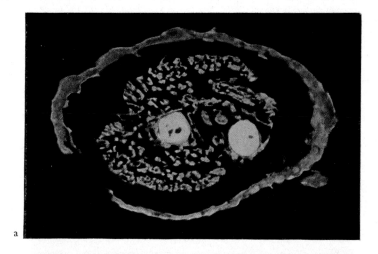

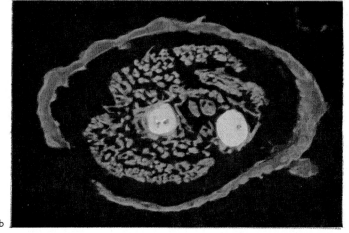

Fig. 159. Newt, cross
section acridine orange.
Transmitted light, UV-free
blue-light excitation.
$M = 80{:}1$ original
enlarged to $M = 200{:}1$:

a Agfacolor UT 16;

b Agfacolor UK 16.

6.976 *Photographic materials*. The finely graduated colours of object detail can only
be rendered satisfactorily on colour film. Reversal film gives better colour rendering
than negative film. Practice has shown that daylight colour film gives better results than
artificial light colour film, since it gives a better rendering of the visual impression,
(see Fig. 159). For black-and-white work, panchromatic emulsions should be used. But
if the image is dominated by green fluorescence colours, panchromatic film may lead
to disappointment when the wavelength of the fluorescent light coincides with the
minimum green-sensitivity of the film. Orthochromatic film will in this case allow
shorter exposures and give better contrast.

6.977 *Special technique*. An extremely sensitive measurement instrument is necessary for the very low level of illumination (CdS photoelectric meter), but this

183

method cannot always guarantee correct exposures as the frequently occurring reddish fluorescence colours give wrong readings, because the film has a different response to those colours. The most reliable (and cheapest) method is making test exposures. For colour photography, black-and-white panchromatic film is used for the test exposures and is correspondingly calibrated with the colour film.

Special applications are described in the literature[58, 59, 60].

6.98   Stereoscopic photomicrography

6.981 Applications. Stereoscopic photomicrography has to represent three-dimensional microscopic objects as three-dimensional photographs. Its main application lies in the low-power range. Special stereo-microscopes are available for visual observation, each for low power magnification. Auxiliary eqipment for the stereoscopic observation at higher magnifications, which are only possible with single-objective microscopes, are no longer available because they were difficult to operate and because the very limited depth of field at high N.A. limited the stereoscopic effect. Basically, stereo photomicrographs can be made with objectives of any numerical aperture, although the use of high-power objectives does not seem advisable, because of their lack of depth of field (see Table 3). Specimens are mainly amplitude objects, and perhaps autofluorescent objects.

6.982 Principles. A stereo pair consists of two individual pictures, two photographs of the same object taken from two different points of view. The centres of the pupils should be considered as such; they act as perspective centres. Their distance $2d$ is known as the separation (Fig. 160). Principal rays from points at different distance from the entrance pupils have a different angle of inclination to each other; in other words, they show parallax. Hence, the pairs of image points belonging to the object points have different abscissae x'.

When the finished individual photographs of a stereo pair are arranged in front of the observer's eyes so that each eye can see the appropriate picture without undue eye strain, since both photographs are observed simultaneously (physiologically) the eyes are induced, because of the different abscissae, to converge the directions of vision towards the image points, as they do when looking directly at object points distributed in space. Therefore, each spatial image point lies at the intersection of the directions of vision for the observer. We can draw three important conclusions from the above—

1. The spatial image is only then "correct in depth" and upright, when the right-hand photograph is observed by the right eye, and the left-hand photograph by the left eye of the observer. Should the two photographs change place, a pseudo-stereoscopic effect is obtained.

2. Since only converging directions of vision are possible when looking at an object in space, the individual photographs should be so separated for viewing, that

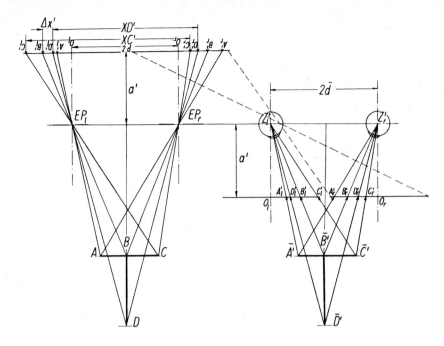

Fig. 160. Principle of stereoscopic photography and viewing.

$ABCD$ object
$A'B'C'D'$ r and l right and left-hand images
$\overline{A}'\overline{B}'\overline{C}'\overline{D}'$ spatial image
XC' like the abscissae (not indicated) XA' and XB'
$XD' < XB'$.

$\Delta x' =$ maximum abscissae difference in one image for the images of two points of the front and rear plane of the object in the plane of symmetry
$Z'r$, $Z'l$ centres of rotation in the observer's eyes.

the directions of vision to the image points farthest away from the observer should still be convergent, or at the most parallel. The farthest away spatial point is then at infinity.

3. The ratio of the width to the depth of the stereoscopically represented object is only then true to the original when the points of rotation Z' of the observer's eyes (or of their virtual images produced in the stereoscopic viewer) are orientated with regard to the individual photographs in exactly the same way as the entrance pupils of the photographic device were orientated when the photograph was made. Only then have the directions of vision the same inclination to each other, when looking at the image points, as the principal rays during exposure. This means that when viewing photographs magnified M times, the viewing distance must be $M \cdot a'$ and the eye distance must be $M \cdot 2d$. In all other cases, the impression of depth would be distorted. Such distortions are often intentionally introduced in stereoscopy for certain reasons, but have little importance in stereoscopic photomicro-

185

graphy, as long as the two eyes are not confronted with hardly comparable aspects of the object by the choice of too great an angle of convergence.

Relaxed viewing of stereoscopic pairs is only possible when the parallax difference occurring in the photograph does not exceed 70′. The parallax is here the angle enclosed by two rays which emerge from two different centres of perspective and both travel to a common point. It is therefore double the convergence angle of a single eye against its normal position when looking at a point situated at a finite distance on the central position normal to the eye distance.

Table 17 gives the parallaxes obtained when looking at points at different distances. Next to the parallaxes are listed the abscissae differences $\Delta x'$ for these angles at a reference distance of 25 cm. The figures in the table indicate that, when looking at different points in space, the eye must be capable of giving parallax differences of approximately 70′, each time, when going from infinity to 10 feet, from 10 to 5 feet, from 5 to 3 feet, and so on. In the case of stereo photographs taken from a distance of 25 cm, and with separation of 6 cm the same distance intervals would yield abscissae differences of 2·5 mm each time in each picture of the stereo-pair, which when viewed from the conventional distance, each time correspond with parallax differences of 70′ for both eyes. The lens separation is considerably reduced when taking extreme close-ups. As a result of this, the useful spatial depth is increased.

It should be mentioned that the abscissae difference of 2·5 mm should be made smaller when the final picture is viewed in a magnifying stereoscope. When Mg_{st} is the magnification of the viewer,

$$\Delta x'_{max} = 2\cdot5 : Mg_{st}\, \text{mm}\,.$$

This principle has been applied in various ways. In order to review the values of the abscissae differences $\Delta x'$ for various arrangements and optical means, a few geometrical observations are given with the explanation of these different applications. The equations will allow the reader, by the substitution of relevant numerical values for each individual case, to obtain the values of $\Delta x'$, and hence the approximate value for the depth in space.

6.9821 Converging axes. For the magnified representation of a near object, it was an obvious solution to have the optical axes of the lens systems converge to the object. This gives the arrangement illustrated in Fig. 161 with two optical axes, both showing an angle of $\pm\varphi$ to the normal NN. Fig. 162 explains the corresponding relations. Corresponding to the axes, the object planes and image planes are also tilted towards each other. Since the depth of field ranges of the two photographs are also tilted towards each other, increasing lateral distance from the image centre will produce different degrees of sharpness in the two photographs.

The depth of field limiting planes are intersected by the optical axis at P_v and P_r, and by the normal of the arrangement at P_v^+ and P_r^+. The distances between these points and the focusing plane EE are projected in the object plane by the principal rays as ΔX_r and ΔX_v.

186

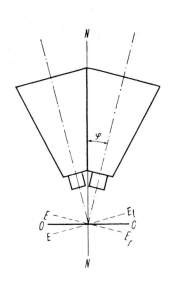

Fig. 161. Arrangement for photography at inclined axes.

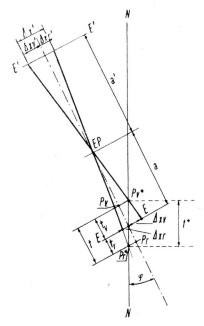

Fig. 162. Determination of the maximum abscissae difference for photography at inclined axes.

The diagram gives an apparent distance of the two points P_v^+ and P_r^+ in the object plane, taking into account that for photomicrophic work it can be taken that $t_r + t_r = \frac{1}{2}t$ (see 1·75)

$$\Delta x = \Delta x_v + \Delta x_r = \frac{t \cdot \mathrm{tg}\,\varphi}{2 - \dfrac{t}{a}} + \frac{t \cdot \mathrm{tg}\,\varphi}{2 + \dfrac{t}{a}}$$

After transition to the object space and introduction of focal length and scale of reproduction instead of a, the lateral distance of the two points in the image plane will be—

$$\Delta x' = M \cdot t \cdot \mathrm{tg}\,\varphi \, \frac{4f^2\,(1+M)^2}{4f^2\,(1+M)^2 - M^2\,t^2} \tag{37}$$

Since the product $M^2\,t^2$ is so small that it can be ignored, we obtain

$$\Delta x' = M \cdot t \cdot \mathrm{tg}\,\varphi \tag{37a}$$

The maximum depth of the object which will be sharply rendered is determined in the direction NN by $t/\cos\varphi$.

With stereo microscopes, the angle of convergence with the normal is approximately 7°, which corresponds to the convergence of each eye when looking at an object at 250 mm distance. The old stereo cameras designed for use with the objectives of stereo microscopes did have this angle, which should not be exceeded without good reason. Working with attachment cameras in conjunction with stereo microscopes offers corresponding possibilities. When two identical attachment cameras are available, living organisms can be photographed in three dimensions, provided that the light source is sufficiently strong.

Another variation of this type of stereo photography is to use a single-lens complete camera for macrophotography, which is tilted in relation to the object. Another method is to maintain the camera in position, but to tilt the object by means of a tilting stage through $\pm \varphi$ against the normal position.

The arrangement with tilting camera, where object and illumination device automatically keep their relative position, can even be used in the case of unstable specimens which would be upset by even a slight change from the horizontal.

6.9822 Parallel axes. Parallel axes position offers the opportunity of arranging the two photographs side by side on a 9×12 cm negative, which is easily and accurately done by means of a multiple-exposure device. Object or camera are then displaced in relation to each other, so that the object appears successively in the two halves of the 9×12 cm plate and both halves show the same image content (Fig. 165). Fig. 163 gives the relations valid for one of the halves. Let us briefly consider the abscissae differences.

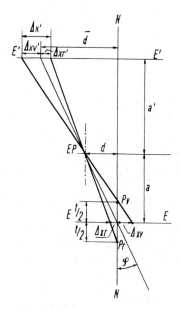

Fig. 163. Determination of the maximum abscissae difference for photography at parallel axes.

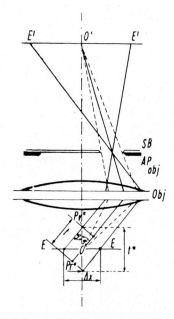

Fig. 164. Stop in the exit pupil of the objective Determination of differential parallax (maximum abscissae difference).

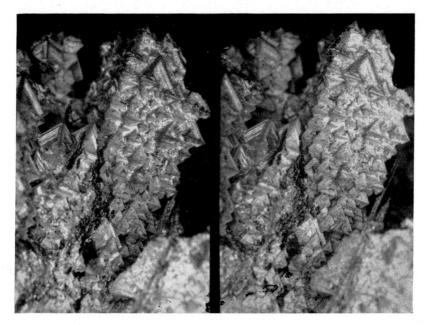

Fig. 165. Corundum. Incident light—dark ground; $M = 2.5:1$.

The depth of field on the normal between P_v and P_r is projected as Δx_v and Δx_r in the focusing plane by the principal rays leading to these points. We have:

$$\Delta x = \Delta x_\mathrm{v} + \Delta x_\mathrm{r} = \frac{d \cdot t}{2a + t} + \frac{d \cdot t}{2a - t}$$

After transition to the image side and introduction of focal length and scale of reproduction, we find:

$$\Delta x' = \frac{\bar{d} \cdot t \cdot M^2}{2f(1 + M)^2 + Mt(1 + M)} + \frac{\bar{d} \cdot t \cdot M^2}{2f(1 + M)^2 - Mt(1 + M)} \tag{38}$$

As $\bar{d}$ represents a quarter of the length-side of the negative material, its value for a 9×12 cm plate is 30 mm. Since already at relatively small scales of reproduction the second addendum in the denominator has a negligible value compared with the first, the equation for rough calculation is:

$$\Delta x' = \frac{\bar{d} \cdot t \cdot M^2}{f(1 + M)^2} \tag{38a}$$

6.9823 Shifting the aperture diaphragm. The above described methods are mainly intended for use in macrophotography. The most frequently used method with the microscope is to shift the aperture diaphragm, keeping the direction of photography constant in relation to the object. Fig. 164 illustrates this principle. An eccentrically

189

placed aperture in the aperture diaphragm (AP) passes an oblique pencil of rays from the total beam of light which can image the object through the objective. It produces one of the images, in this case, the right-hand image. After rotating through 180°, the eccentric aperture comes on the opposite side of the optical axis, and the second image is formed. In the case of aplanatic systems, such as microscope objectives, the heights of the imageforming rays in the exit pupils have the same ratio to each other as their numerical apertures on the side of the object. The effective numerical aperture of the rays passing through the eccentric aperture can therefore be defined by the ratio of their diameter to the diameter of the exit pupil. We must further assume that the edges of the exit pupil and the eccentric aperture touch each other at one point. The angle of inclination of the principal ray of the image-forming pencil of rays is then:

$$\sin \varphi = \frac{N.A._{obj} - N.A._{eff}}{n} = \frac{\Delta N.A.}{n}$$

where n is the refractive index of an immersion medium possibly used. In the case of dry objectives, $n = 1$. Depth of field and useful scale of reproduction depend on $N.A._{eff}$. The depth of field is then [according to equation (32)]:

$$t_2 = \frac{n}{7 \cdot M \cdot N.A._{eff}} \; ; \; M \text{ being } 500 \, N.A._{eff}.$$

The eccentric aperture should have such a diameter that $N.A._{eff} \leq \frac{1}{2} N.A._{obj}$.

The two limits of the depth of field are perpendicular to the principal ray of the image-forming pencil of rays, and intersect the optical axis at P_v^+ and $P_r^{\div}$. The depth t^+ in the direction of the optical axis is therefore—

$$t^+ = \frac{t_2}{\cos \varphi} \tag{39}$$

and the projection of t^+ on the focusing plane EE—

$$\Delta x = t^+ \cdot \text{tg}\, \varphi = t_2 \cdot \frac{\text{tg}\, \varphi}{\cos \varphi}$$

By substituting the previously obtained values for these quantities, and transferring to the image side, we have the geometrical-optical equation—

$$\Delta x' = \frac{\Delta N.A. \cdot n^2}{7 \cdot M \cdot N.A._{eff} (n^2 - \Delta N.A.^2)} \tag{40}$$

These considerations are not intended to realize exact measurements of the obtained depth of field and abscissae differences, but lead to the following deductions:

1. In principle, any objective can be used to obtain a stereoscopic effect.

2. This effect lies within the range of possible vision by the pair of eyes, as can be easily demonstrated with numerical examples.

190

3. When the aperture is placed in the image space, the eccentric aperture will determine the effective numerical aperture, and only this is decisive for the useful scale of reproduction and depth of field in the direction of the image-forming pencil of rays in the object space. The diameter of the eccentric aperture should therefore never be smaller than necessary for the depth of field required. On the other hand, with regard to the stereoscopic effect, the diameter of the diaphragm aperture should not exceed half the diameter of the pupils.

4. The principal rays of the image-forming pencil of rays show a strong inclination towards the optical axis in the object space, and when viewing the finished photograph, this angle is very small. This means that the space is represented in a strongly distorted manner, which can lead to difficulties in the conception of the stereoscopic image, especially in the case of object details, which mainly stretch in the direction between the optical axis and the two principal rays of the image-forming pencil of rays. These details are observed simultaneously from left and right. Yet even this kind of photograph can be useful to demonstrate the position of details.

5. Stereo photomicrography is often made impossible in high-power work by the extremely small depth of field.

6.983   Stereoscopic macrophotography

6.9831 Suitable specimens. The most suitable specimens are those which can be illuminated by reflected light, and whose extension in depth does not exceed the area of depth of field. It is possible to use transmitted light for transparent specimens. The use of a ground-glass or opal disc is recommended to ensure perfectly even illumination of both individual photographs.

6.9832 Converging axes. There is no special equipment available for this method at the present time. Nevertheless, this technique can be useful in conjunction with small-format cameras, and can be improvised in a relatively easy way. Single-lens reflex cameras with extension tubes or bellows are ideal for this purpose. There are two methods of operation. The specimen is placed on a tilting stage and two photographs are taken in succession at an angle of $\pm 7°$ with the horizontal, or the specimen is shifted so far to the left or right (while remaining in the same horizontal plane) that the camera has to be tilted through a corresponding angle to make the same position of the specimen appear in the negative frame.

The second method is the more difficult, since one works mainly with an upright format, which precludes using the base adjustment screw. Independent of the method for changing between exposures, the specimen should on no account be tilted or its height in the image plane changed. Should a horizontal photograph be made (which is usually not recommended because of the less efficient utilization of the negative format) and if miniature or roll-film is being used, the sequence of photographing

191

should be so chosen that the photographs and the film go in the same direction. When the film is advanced from left to right, the left photograph is taken first, then the right. The photographs will then be correctly orientated after printing. This method is obviously advantageous for $2\frac{1}{4} \times 2\frac{1}{4}$ in. film, particularly when the stereoscope is large enough to take two pictures together at the correct separation. When the illumination gives harsh shadows, these should not change their position between the two exposures, otherwise they would be seen as lying in a different plane. It is therefore essential to move the illumination device simultaneously with the specimen. But it is best to avoid harsh shadows, and ensure a uniform illumination from all sides if possible, which also penetrates into the deeper parts of the specimen (reflected light). Seen in the normal way, such results would appear uniform and devoid of contrast; seen in a stereoscopic viewer, this is no longer so. This type of even illumination also brings out fine detail.

Scale of reproduction and diaphragm adjustment are so chosen that the object is not too strongly magnified when observed in a stereoscope with a magnification Mg. Apart from this, the obtaining of the required depth of field is more important than entirely filling the negative format.

The prints are mounted on card for viewing in the stereoscope. When modern viewers are used, the two photographs should be mounted at such a distance from each other, that corresponding points of the farthest object detail are as separated as the eye separation of the observer. It is even better to make the print separation slightly shorter. The photograph of the specimen from the right-hand side yields the right-eye viewer picture, and vice versa. It should be noted that each individual photograph will be upside down when exposed, and therefore should be reversed for viewing (transposing the two pictures each for itself) in case it was not possible to apply the film transport technique described above.

6.9833 Parallel axes with no tilting of the specimen. This is an excellent method for the vertical 9×12 cm camera, since the two pictures, each measuring approximately 55×90 mm, can be accommodated on a single plate. As the camera position is fixed, the specimen is displaced between exposures along the length-side of the negative format so that the same specimen is photographed in each case. This displacement can be effected by means of a mechanical stage or the edge of a carefully arranged ruler. The lengths of the displacement are determined by test, and observation of the focusing screen image. The nonexposed negative half should be covered by a plate in front of the negative material. Since during exposure, each photograph is reversed, the pair is wrongly orientated on the plate. Stereo photographs with a vertical camera are therefore best made with a multiple-exposure device, because it is then possible to interchange the relative positions of the two photographs. This eliminates subsequent transposing, and prints can be directly observed in the stereoscope, without any cutting up.

The technique for stereo-macrophotography with the standard vertical camera and multiple-exposure device, is as follows—

1. Provisional focusing of the specimen by using one half of the negative format. This can be marked on the ground-glass screen or a mark placed in the multiple-exposure device.

2. Preparing the specimen displacement (lateral mechanical stage or ruler), parallel to the length-side of the negative format. The specimen must be moved in the direction of the long side of the format—no change must occur in its height.

3. Determining the length of the displacement. At the final position for exposure, the two images of the farthest away object point should be separated by approximately 60 mm.

4. The specimen is now placed in the in-between position; the focusing plane is brought in the average object height by maximum lens objective aperture; the aperture is adjusted, taking into account the depth of field and the resolving power, and the exposure time is determined.

5. The specimen is placed in the left exposure position, and the left half of the photographic material in the camera is covered up. The multiple-exposure device is slid to the right as far as it will go, so that the exposure is made on the left half of the plate. The plate holder is inserted.

6. The slide is removed from the plate holder. The exposure is made, the plate holder closed and removed.

7. Specimen, plate cover and multiple-exposure device are all placed in the opposite positions. The plate holder is inserted.

8. The slide is removed from the plate holder. The exposure is made, the plate holder closed and removed.

With regard to illumination, choice of scale of reproduction and diaphragm adjustment, the observations made in 6.9832, all apply.

6.984 Stereoscopic photomicrography with the stereo-microscope. This type of photograph is made with convergent axes (see 6.9821), or the arrangement is basically the same as for the microscope with one objective; the choice depends on the design of the microscope. Since special cameras are not available for this purpose, an attachment camera or miniature camera with microadapter is used. A photo sliding tube is available, which is successively clamped in each of its end positions. The photo sliding tube also contains an iris diaphragm which is closed in order to increase the depth of field. If no photo sliding tube is available, this diaphragm will not be present, so that the depth of field cannot be influenced. In this case it will be necessary to make certain before exposure that the two images are sufficiently sharp in the most important image points. Finally, the use of a green filter is recommended.

6.9851 Suitable specimens. This subject has been anticipated in 6·9823, points 4 and 5. Basically, such photography is possible both with reflected and transmitted illumination, bright-field as well as dark-ground. Very shiny specimens are less suitable for reflected light, particularly in bright-field; because of the different inclinations of the surfaces and the arrangement which is usual in bright field there is a danger that large areas will show different brightnesses in the two photographs of the stereo pair (see below). With transmitted illumination, specimens with many phase components may cause illumination difficulties.

6.9852 Equipment. For work in bright field, the microscope should have a laterally adjustable aperture diaphragm in the illumination device. There are no special requirements with regard to the camera, although an attachment camera cannot be used for dark-ground photography (see 6.9853). The 9×12 cm bellows camera has the most universal application possibilities, so that we shall now mainly concentrate on this type. As the used range of scales of reproduction should lie below the useful scale of the objective (secondary magnification in the stereoscope), particular attention should be paid to sufficient flatness of the image field by objectives with high numerical aperture. Plano-objectives are therefore preferred.

6.9853 Illumination; adjusting the diaphragm. The eccentric aperture is positioned in the exit pupil of the microscope when dark-ground illumination is used. It is therefore placed on the eyepiece or projection eyepiece. The simplest form of eccentric aperture is a stop with a straight edge, which is so arranged in the plane of the exit pupil above the eyepiece that it covers an identical area of the exit pupil in any position. The easiest way of adjustment is to rotate the stop together with the eyepiece. The use of a very low power eyepiece is recommended.

When bright-field illumination is used, the far easier possibility of stopping down in the illumination space by means of a laterally adjustable aperture diaphragm is utilized. Since only illumination rays, and not the light refracted at the object are taken in by the diaphragm, higher resolving power results, as well as a naturally lesser depth of field than obtained by arranging the stopping down device in the image-forming space. A scale of reproduction of higher than $500N.A._{\mathrm{eff}}$ can be envisaged, but too high magnification should be warned against as this would impair image quality.

Since the illumination cone lies on one side in the refracted light used for image formation, the phase components of the specimen will be differently represented in the two photographs of a stereo pair. The phase components should occupy only a minor part of the bright-field stereo photograph.

6.9854 Practical hints. The same rule is valid for both the eccentric aperture above the eyepiece and the position of the aperture diaphragm of the illumination device, namely that the right-hand photograph is produced when the diaphragm passes the light beam on the left of the optical axis, and vice versa. The microscope with eye pieces or positive projection eyepieces produces upright pictures, which lie in the same position in the camera. With a 9×12 cm camera, it is useful to employ a central area of about 6×9 cm, and to insert the corresponding format-limiting mask belonging to the multiple-exposure device or to the camera, below the image plane. When working with a multiple-exposure device, the right-hand picture will be produced on the right-hand side of the plate, and conversely. Singly taken photographs should not be exchanged nor reversed. Rollfilm facilitates correct orientation, when the sequence of exposures and the direction of the film transport are opposed to each other. When working with homal projection eyepieces the individual pictures are inverted; therefore they must be turned (each for itself) when mounting. The multiple-exposure device must be so used that the right-hand picture comes on the left-hand side of the plate and vice versa. The individual photographs are properly orientated on rollfilm when the exposure sequence and film transport go in the same direction.

6.99   Photomicrography with electronic flash

6.991 Applications. Correct photomicrography with electronic flash demands relatively high expenditure on apparatus and, as long as no electronic flash equipment is designed specifically for microscope work, a certain skill in manual work on the part of the operator. Electronic flash equipment, specifically designed for photomicrography would consist of a micro-flash, electronic flash generator and a special lamp housing, as well as a low-voltage lamp for focusing purposes with its own mirror housing fitted with iris diaphragm (if not already available). Electronic flash is particularly recommended for the photography of living organisms, which would be damaged by continuous intense illumination. It is, however, often possible to avoid these unfavourable influences by appropriate filtering of the light (UV or infra-red barrier filters, limitation to narrow wave-bands). Low-voltage lamps are often sufficient for the supply of the intense illumination necessary for short exposure times.

6.992 Set-up. If the above described special electronic flash equipment is not available, any ordinary electronic flash device may be used. The flash tube (without reflector) is arranged in the best position in the illumination space of the normal photomicrographic apparatus for a uniform illumination of the object field, namely the focal plane of the microscope condenser, in the proximity of which lies the aperture diaphragm. The normal microscope lamp is used for focusing and visual observation. When the flash tube does not have a matt glass envelope, the normal light beam will not lose much of its intensity by the presence of the flash tube. This arrangement will prove sufficient in many cases, but does not represent the apex of performance or picture quality, apart from being uneconomical. A disadvantage is that this technique does not allow the limitation of the illuminated area to that being recorded on the photographic material because the iris diaphragm of the microscope lamp will not operate with the electronic flash. The results are frequent over-illumination and reduced picture contrast. Non-uniform illumination can also easily occur with high-power work, because the flash tube never lies exactly in the focal plane of the condenser. In the case of condensers with a short focal length the flash tube cannot be arranged closely enough to the condenser. Another disadvantage is that the depth of field suffers through the dimensions of the flash tube in relation to the focal length of the condenser. As these dimensions frequently do not harmonize with the effective aperture of the condenser, a very large part of the flash will not be used. Optimum utilization of the flash, combined with optimum picture quality, is obtained in the specially designed electronic flash equipment, based on the Köhler illumination system. The dimensions of the bull's-eye condenser depend on the size of the flash tube.

$$\text{Diameter}_{\text{flash tube}} \cdot \text{Aperture}_{\text{bull's-eye}} = \text{Diameter}_{\text{object field}} \cdot \text{Aperture}_{\text{condensor}}$$

$$\text{Diameter}_{\text{bull's-eye}} = \text{Diameter}_{\text{object field}} : M_{\text{condenser}}.$$

An auxiliary light source is required for focusing and observation of the specimen. The flash tube itself can be used as such, when intermittent illumination, with a frequency of usually 30 cps, is possible. If this is not so, a microscope lamp can be used as an auxiliary light source. The light source itself can be imaged through the transparent flash tube, or can be reflected from the side. This last technique requires two complete microscope lamps; one for the flash tube and one for the projection lamp. They should be so arranged that their field diaphragms are optical conjugates. The auxiliary light source is reflected by a 45° plane-parallel glass, which should have a high transmission and a low reflection factor. This condition is fulfilled by ordinary glass ($\tau \cong 0{\cdot}92$, $P \cong 0{\cdot}08$). There is some possibility that the auxiliary light source, which keeps burning during the flash exposure, can exercise an undesired influence on the image formation. This requires adjustment possibility of the brightness of the auxiliary light source, otherwise the synchronized focal plane shutter, set to $\frac{1}{20} - \frac{1}{50}$ second may cause feeble double contours.

6.993 Exposure determination. This is more difficult with electronic flash than with other light sources. The brightness in the film plane can vary in a ratio of as much as 1 : 600 because of changes in the illumination aperture, absorption conditions of the specimens, and filter factors. This variation cannot be compensated by altering the exposure time since the flash duration is constant with most electronic flash equipment. The development latitude of negative materials is equally incapable of absorbing such a range, if only one type plate or film must be used. Satisfactory results are obtained when the speed of the negative material is varied in relation to the scale of reproduction, and within the appropriate limits set by development.

196

A practical test with a commercial electronic flash equipment[61] gave the following results—

M	Film speed
< 25:1	8 ASA + neutral density filter
25:1—150:1	8 ASA
150:1—400:1	40 ASA
>400:1	100 ASA

This technique has some drawbacks. The variation in development and the continuous change of film type are not only irrational, but also make printing more difficult because of differences in gradation.

A better method is to expose the negative material by constant development conditions for the most advantageous density and gamma, and to regulate the flash intensity with a set of graduated neutral density filters, or better still, with an illumination regulator[62], consisting of two polarizing filters so that the same brightness is always present in the focal plane. This regulation can be checked with a light-measurement device which is calibrated in the auxiliary light source. This calibration is effected as follows: a series of flash exposures at different position of the illumination regulator (or with different neutral density filters) is made of a test object. After processing of the test strip, the illumination regulator is set to the value which gave optimum density with flash. With the auxiliary light source, a photo-cell placed in the focal plane or a conjugate position, and a measurement instrument of appropriate sensitivity, a photo-current of a certain value is generated, which is recorded definitively. For all other exposures under the same conditions of film material, development and a continuous burning auxiliary light source, the illumination regulator is actuated for as long as the needle of the measurement instrument indicates the calibration value. The adjustment of the auxiliary light source in relation to the flash tube is of prime importance. Its field diaphragm should also be adjusted to the opening equivalent to the diaphragm of the flash housing.

But even this technique will require the use of negative materials of different speeds. The difference, however, is that they can be developed to the most favourable density and constant gamma.

Some electronic flash equipment has variable flash intensity, which obviates the need for the illumination regulator. The calibration against the auxiliary light source is basically as described above, but this time not by altering the adjustment of the illumination regulator, but by varying the flash energy. Should the range not be sufficient, the illumination regulator will have to be used as well. A calibration curve will result when the brightnesses obtained with the auxiliary light source (measured in scale-graduations of the measuring instrument) are plotted against the correspondingly required flash intensities.

6.994 Camera. As the main application of electronic flash is in the field of the photography of living organisms in different stages of development, the miniature camera is the most rational choice. This camera should preferably be used with an attachment camera which allows continuous observation of the specimen, even during exposure. Synchronization and release of the flash is obtained by using the X-socket built in the camera.

6.995 *Filters and photographic materials.* The high blue-content of flash light leads to pronounced flat gradation, even in the case of rather hard-working negative materials and developers. Although a good definition of all object details may be desirable, a filter will nevertheless be used to obtain higher contrast. A yellow-green filter, with an acceptably low filter factor, will be appropriate. This factor must of course be taken into account when selecting the speed of the negative material.

Orthochromatic film is nearly always sufficient for black-and-white photography. Colour film should be of the daylight type.

VII. THE EXPOSURE RECORD

7.1 Routine

The significant data of each exposure should always be recorded immediately after it has been made. These records will only be of value when they give all the required information and can be easily and quickly located. The data should therefore be entered in an exposure day book, or better, entered on printed index cards. It is a good idea to have a space on the cards for a print of the photomicrograph.

The many varieties of photomicrographs should be classified under different main groups and sub-headings. The type of system will depend on the character of the

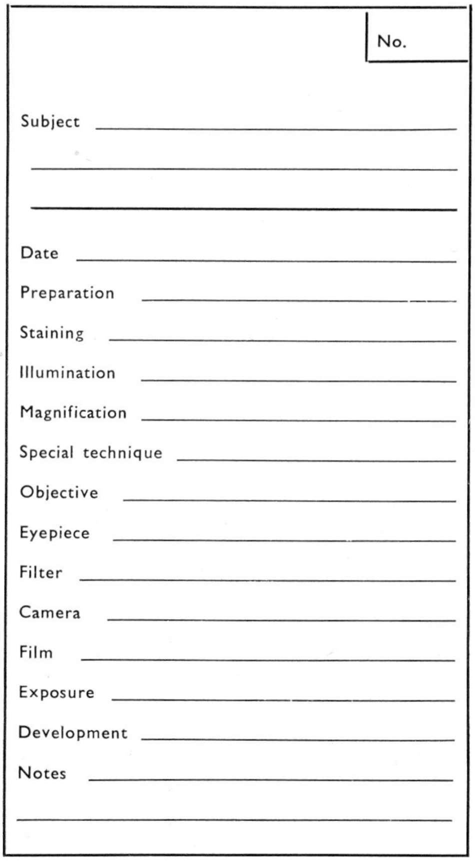

Fig. 166. Index card for photomicrographs.

microphotographic laboratory. When the work is limited to research in special spheres, the main groups can be given the names of those fields, e.g. histology, cytology, pathology, bacteriology, etc. An appropriate sub-division of these main groups will lead to further specialization, while the photographs can also be divided into the different techniques, such as bright-field, dark-ground, phase-contrast, etc. When a great variety of work is undertaken, a basic division of different techniques may be the more efficient.

The day-book or index card should always contain the following headings, without which the photomicrogram would be useless—

1. Name of object
2. Origin of object
3. Preparation method
4. Type of illumination
5. Scale of reproduction

Beyond this basic information, the records should give data which gives the operator guidance for the photography of similar objects in the future, and so avoid time-consuming tests for optimum picture quality. These points comprise—

6. Microscope stand
7. Objective
8. Projection system
9. Condenser
10. Setting of aperture diaphragm
11. Light source and type of lamp
12. Adjustment of field diaphragm
13. Filter
14. Camera
15. Photographic material
16. Exposure time
17. Developer (name and dilution)
18. Development (time and temperature)
19. Reference number

The reference number should be written in ink on the emulsion side of the plate or film.

7.2 Storage of negatives

Negatives should be stored so that a first-class contrast print or enlargement can be made at any time. The negatives must remain undamaged in the archive. Plates are best stored in metal cabinets or index cabinets, each in a transparent envelope. The negatives can also be stored in the empty plate boxes; the print accompanying each plate will protect it against damage.

Miniature films are best kept in the well-known file boxes for 35 mm film storage and indexing. Each box carries 100 cards each taking a strip of six negatives in translucent envelopes. The cards are numbered for easy reference, while an index for 600 negatives is kept in a pocket in the lid. The space for data on the index sheet is insufficient for our purposes so that such a filing system should be used in conjunction with the above-mentioned card index system. The advantage of this system is that negatives can be selected without removing them from their protective envelope. Films should never be stored in roll form, as this invariably leads to scratches and other damage.

200

VIII. PRACTICAL HINTS

8.1 For exposure

8.11 Increasing contrast with diffuse-reflecting and refracting objects. Microscope and photomicrography work with soft and diffuse reflecting objects (e.g. polished sections of coals, ores and some metals) needs an increase of contrast. Adjustment of the aperture diaphragm within the permissible range does not give a sufficient increase, because the diffuse reflection from all parts of the illuminated field causes the rays to be reflected in all directions, and thus to fog the image of the object. An increase in contrast can be obtained by reducing the illuminated field, i.e. narrowing the field diaphragm. Only then will the structural elements of the object show a considerable contrast (Fig. 167).

The same effect, although not so pronounced, is observed with high-power transmitted light bright-field work. Here too, a noticeable increase in contrast—and thereby a better rendering of fine detail—can be obtained by stopping down the field diaphragm to a minimum value. Colour film demands the use of achromatic condensers in this case. Should ordinary bright-field condensers be used, the field

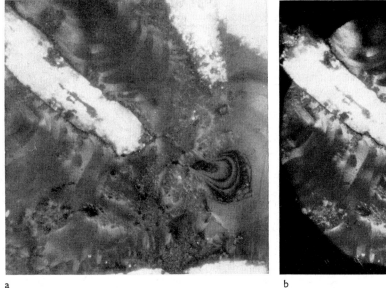
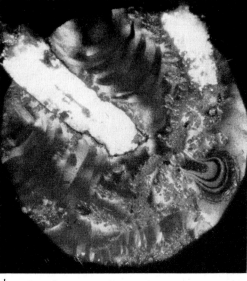

a   b

Fig. 167. Contrast increase by means of field diaphragm. Psilomelan from Gehege near Brotterode. Incident light—bright field, polarizers +; M = 80:1; *a* (left) field diaphragm image > negative size, *b* (right) field diaphragm closed.

diaphragm would be imaged with coloured fringes, which at a small field size, would merge to a coloured field. The photograph would then show a colour cast, the colour of which varies with the height adjustment of the condenser. It is always very difficult to determine the exact image position of the field diaphragm in the case of chromatically non-corrected condensers. This makes it almost impossible to make colour photomicrographs suitable for reproduction.

8.12 Avoiding and eliminating objectionable reflections

8.121 Reflections off the specimen (non-metallic surfaces). Reflections cannot always be avoided with objects for macrophotography having convex or polyhedron form, shiny or half-matt surfaces. Their reflections are sometimes desirable, as they demonstrate the type of surface and heighten the threedimensional effect. But they are objectionable when the photograph should show a structure beneath the surface, or the best rendering of the design or differences in colour of the material produced by pigments. Polarized light can be of value. It eliminates surface reflections, or in any case reduces them to an acceptable level. A single polarizing filter is often sufficient, positioned before the light source or the objective. The direction of vibration, marked on the filter mount, is so adjusted that it coincides with the plane of the camera axis and the direction of incidental light. The reflections can be reduced to a minimum by rotating the filter and changing the angle of incidence of the light, while keeping the direction (azimuth) constant. With crossed polarizers before both microscope lamp and objective, reflections can be eliminated at any angle of incidence, which leaves the operator free to position the light source most favourably for other aspects of picture quality.

Surface reflections from humid or slimy biological or pathological specimens can be eliminated by embedding the object in a liquid.

8.122 Reflections off convex or rough metallic surfaces in macrophotography. This type of reflection cannot be avoided. It is therefore recommended to use them for picture quality by appropriate arrangement of the light source. In this way, cleavages or crystalline formations should not be directly illuminated by a light source with a limited field aperture, but by means of a diffusely glowing surface with a sufficiently large azimuth of illumination. This avoids the highlights produced on fractions by direct illumination, which lead by stronger stopping down of the objective to the formation of pronounced refraction discs, so that the production of dark shadows in their immediate proximity is avoided. With convex metallic surfaces, the part of the surface illuminated in bright-field can be enlarged in similar fashion.

8.123 Elimination of reflection in the lenses of objectives for reflected light microscopes. Reflections will result by any transition of the light from one optical medium to another. They can be strongly reduced by a coating on the lens surfaces, but cannot be entirely ignored in practice. They are particularly evident when it is not possible

202

entirely to separate the lenses in both the illumination and image spaces from the optical path in the objective. This is so with reflected-light microscopes. A glance in the eyepiece before placing the specimen on the stage will give the required information on the intensity of reflected light and its distribution in the field. This reflected light can be influenced by adjusting the diaphragms. The distribution of reflected light should be uniform over the entire field when the exit pupil of the objective is fully illuminated.

By the observation of regular and strongly reflecting specimens such as metal sections, the internal lens reflections are of no importance, since their intensity is too small compared with the light reflected from the object. But this relation is different for the study of dark and more or less matt objects. The small reflection is here joined by a general light scatter, so that only a small fraction of the light falling on the specimen contributes to observation or photography. With Köhler's illumination, the reflected light, independent of the brightness of illuminated parts of the specimen, will become less with decreasing narrowness of the field diaphragm. This gives the following rule for highest possible contrast: The aperture diaphragm should be opened just enough to allow the exit pupil of the objective to be fully illuminated, but not more, as otherwise scatter would be produced by reflection from the mount of the objective. The field aperture should always be closed as far as possible, if necessary at the cost of the negative format area.

Reflections produced by coated objectives are mostly purple or bluish in colour, since, strictly speaking, decrease in reflection is only valid for green light. A very strong green filter will remedy this defect.

Lens reflections with low-power reflected light objectives can, should they still be present after observation of the above hints, be eliminated in the following manner[63]. Crossed polarizers are arranged at 90° (polarizer in the illumination space, and analyser in the image space). A mica quarter-wave plate is inserted between specimen and objective. The principal vibration directions of this plate should be at an angle of 45° to the vibration directions of polarizer and analyser. Since the rays pass twice through the mica plate, a linear polarized vibration will once more result, which is, however, rotated through 90° against the vibration direction of the polarizer. This results in the vibration of the image forming rays passing through the objective to be parallel to the vibration planes of the analyser and thus being transmitted by the latter. The rays reflected by the lens surfaces, however, which after all produce the reflection, retain their plane of vibration and are absorbed by the analyser (Fig. 168).

This technique is naturally not suitable for work in polarized light between crossed polarizers, and is unnecessary in this case since lens reflections are absorbed of themselves.

8.13 Avoiding scatter with attachment cameras. We have already mentioned in section 5 that subdued lighting in the laboratory is necessary for optimum focusing. This is particularly valid with attachment cameras fitted with beam splitter. Here is a danger of reflection when the focusing system would be pointed at a sunlit window or

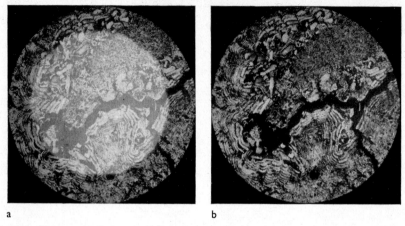

a   b

Fig. 168. Elimination of objective reflections. Lignite. Incident light—bright field; $M = 63:1$
a without, b with, polarizers and $^1/_4 \lambda$ plate. Both photographs were made with the same test objective, selected to show the strength of the effect. The diaphragm image is a reflection of the aperture diaphragm, and *not* the image of the field diaphragm (see Fig. 167 b).

otherwise strongly illuminated surfaces. For light would penetrate from there into the focusing system, which would be reflected by the beamsplitter in the direction of the projection eyepiece, and from the latter's surface to the film plane, resulting in fogged photographs. It is even frequently possible to distinguish the shape of a window frame or suchlike (Fig. 169). These phenomena will always occur under the given circumstances and will be more noticeable, the longer the exposure time.

The opening of the focusing system should therefore be closed during exposure, by placing a lens cap over the focusing system after final focusing. Should it be

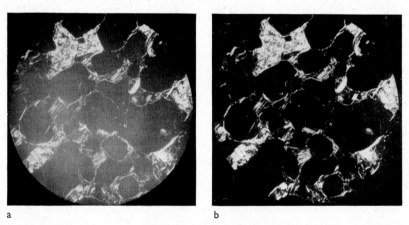

a   b

Fig. 169. Scatter in a camera attachment with dividing cubes. Dunite with chromite from Kraubat/Steiermark. Transmitted light—bright field, polarizers $+$; $M = 20:1$
a Focussing system directed towards window; b Focussing system covered up.

204

necessary to observe the specimen during exposure, an eye-shade should be used which excludes all extraneous light.

8.14 The need for cleanliness. Picture quality can only be obtained when all lenses are clean and dust-free. It is not always possible to make the laboratory dust-free, but the equipment should be sufficiently protected. Transparent plastic covers are recommended. Objectives and eyepieces should be stored away in cases and put back immediately after use. Particularly sensitive systems, whose position in the path of rays tends to image any uncleanliness or dust particle on the photographic material, should be cleaned before use. This concerns mainly the eye lens of the eyepiece or projection eyepiece, especially those of high power (Fig. 170). Cleaning can be effected with a very soft brush or a hand blower, for the careful removal of dust particles. Lens surfaces are best cleaned with special lens-cleaning tissue. Brushes can be cleaned in ether. A warning must be made against the use of a strong blower on the entire instrument, since this would unavoidably result in the penetration of fine dust particles on to internal lens surfaces. Such an action would lead to expensive repairs. There is, however, nothing wrong with the use of a suction device.

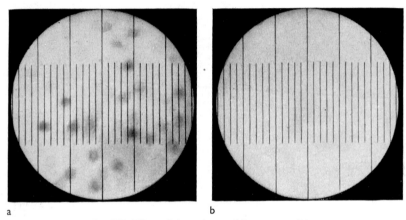

a   b

Fig. 170. Effect of dusty lenses. Micrometer slide:
a dust on projection eyepiece; *b* clean projection eyepiece.

8.15 Pictures for comparison. Comparison pictures demonstrate the difference or identity of different objects in photomicrographic practice. They are also used to show the differences between various optical and preparatory techniques. A similar task is the making of photographs of different sections of a specimen (Fig. 171). When it is not necessary to show the effect of a difference in the optical arrangement, it will be best to keep all imaging and processing conditions constant. Special comparison microscopes with cameras are available, in which the two specimens to be compared are observed and photographed through two objectives and a common eyepiece. If such an installation is not available, one must be content with successive

Fig. 171. Cedar.
Three sections perpendicular to each other. Transmitted light—bright field; M = 63:1.

photography using a single instrument. It is, however, desirable to photograph the two images side by side on the same plate or film, so that they can be printed on the same piece of paper. The multiple-exposure device can be used for plates or cut films. The central portion of the field is used for each exposure. The multiple exposure device for 9×12 cm allows for two pictures each of 55 mm width, or seven pictures of 13 mm width side by side. The last procedure is recommended for narrow objects, such as textile threads. Special adapters can accommodate three 24×36 mm images, or five strips of 11 mm width on a 6.5×9 cm plate.

The miniature camera is the best for a large series of comparison photographs. Standardization is particularly important in this case.

8.16 Including graticules for measurements. It is often advantageous to include a graticule with known graduations for the size determination of parts of a specimen. Different methods are available. The simplest is to use an eyepiece micrometer with the microscope. Since the eye lens or the entire optical system is focused in the area of the micrometer, it is also possible, by appropriate focusing, to obtain a sharp rendering of the graduations in the film plane (Fig. 172). It is best first to focus the micrometer, and then the specimen.

A disadvantage of this method is that the imaged graduations are only a relative measurement for the size of the object parts, since the scale value varies with the objective. Therefore, in each case it should be determined and indicated which distance corresponds with one graduation interval of the micrometer on the object plane.

Fig. 172. Mixed threads
(somolana).
Transmitted light—bright
field; $M = 160:1$.
1 scale interval $= 10\,\mu$m
in the object.

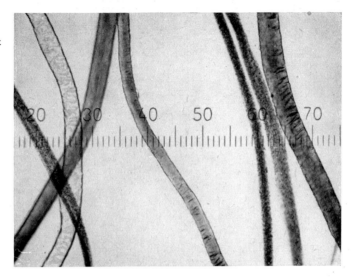

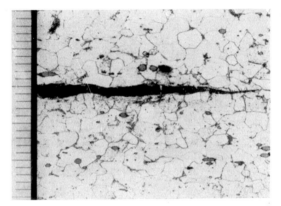

Fig. 173. Steel with slag line.
Incident light—bright field; $M = 200:1$.
1 scale interval $= 10\,\mu$m.

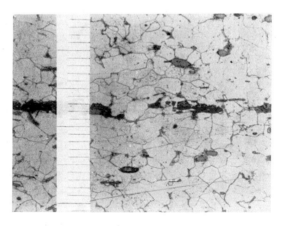

Fig. 174. As Fig. 173.

Another and better method consists in the photographing of a object-micrometer with known graduation values. In this case, the graduations can be recorded at the edge of the photomicrograph, separated from the microscopic picture (Fig. 173), or arranged at the very spot of the field which must be measured (Fig. 174). This is done by first making the usual photomicrograph of the specimen, and then replacing the specimen by the graticule object plate. A second exposure is then made with identical optical arrangement and camera extension. When it is required to have the graticule at the edge of the photograph, this area on the plate or film should be covered during exposure for the specimen. During the second exposure, the other part is covered. The multiple exposure device is very suitable for this purpose[64]. If, however, the graticule should be reproduced within the field of the specimen (Fig. 174), a careful harmonizing of the two exposure times is important. In the case of dark specimens, a negative graticule with bright graduations on a dark background has proved to be more suitable.

Finally, we should refer to the possibility of including a photographed graticule in the photomicrograph during printing, especially with miniature work. A few exposures are made of a specimen and of a graticule (under identical optical conditions), and the exposure times are determined so that the contrast between object and graduations is clearly visible in the print. The negative of the graticule is then enlarged together with the negative of the specimen, or two successive exposures are made with the enlarger (one with the specimen and one with the graticule negative). In this case, too, a negative graticule is used for dark specimens, and a positive for bright specimens.

8.2 Photographic hints

General processing technique of photographic materials will not be discussed in this book, but we would like to mention a few points which will considerably help towards obtaining consistently good results.

1. Always use fresh developer. This is the only way to guarantee the best possible results with an exposure meter.

2. Always use a larger dish than strictly necessary for the size of the plate or film developed, and a generous quantity of developer.

3. The use of a wetting agent in the last wash water prevents drying marks, whose subsequent removal leads to scratches.

4. Intensification of an under-exposed negative is not much use. A new exposure should be made. The same applies in a lesser extent to over-exposure. A reducer can yield acceptable print quality. It is useful to employ a reducer which also gives a finer grain. The negative is first bleached in:

Copper sulphate	100 g
Sodium chloride	100 g
Sulphuric acid (conc.)	25 cc
Water to make	1,000 cc

After rinsing for a few minutes, the negative is re-developed for 2 minutes in:

Para-phenylene diamine	3 g
Sodium sulphite (anh.)	20 g
Water to make	1,000 cc

At this stage, the negative does not show much difference from its appearance before treatment. Immersion in an acid fixer will brighten it quickly. The process may be repeated.

5. Low-level brightness may lead to unacceptably long exposure times. If it is not possible to use a faster material (e.g. infra-red material) previous hypersensitization can help. An increase of about 66% can be obtained by washing the plate or film in water for 5—10 minutes. The following formula gives even higher increase in speed:

Potassium carbonate	1 g
Ammonia solution (0·91)	3 cc
Water to make	500 cc

Immersion during 3—5 minutes; hypersensitized plates only keep for 24 hours.

6. Prolonged development can also shorten exposure by about 8 times. This method is nowadays only used for fluorescence photomicrography, as electronic flash can be used with other techniques. Films of 40—80 ASA, for example, can be developed for 2 hours in Rodinal 1:40 at 18 °C. The gamma obtained is about 2·3, so that the method is only suitable for objects with a very short brightness range. Development fog and grain are within acceptable limits, and picture quality is not impaired to any extent[65,66].

7. Glossy paper is preferable for contact prints and enlargements. Bromide (enlarging) paper only is to be used (also for contact-printing), because this has the widest tone scale.

APPENDIX

TECHNICAL TERMS

Absorption bands. When white light is transmitted through an absorbent substance and causes a reduction in light over one or more wavelength bands, the gaps in the continuous spectrum are termed absorption bands.

Accommodation. The adjustment of the eye at various object distances by changing the focal length of the eye lens.

Actual resolution. Grabner's concept for the efficiency of a microscope by eyepiece observation. The actual resolution Y is the smallest distance between two object points which can just be seen as separate entities. Related to an angle of vision of $4\frac{1}{2}'$ (sharp and comfortable vision)

$$Y = \frac{0 \cdot 32 \text{ mm}}{M g_{\text{microscope}}}$$

Adaptation. The sensitivity adjustment of the eye according to different levels of brightness.

Anisotropic effect. The change in light intensity when rotating the illuminated object between crossed polarizers. The effect varies characteristically with the wavelength of the incident light.

Aspherical. Curved surface, not having the shape of a sphere.

Colour sensitization. The silver halide emulsion of a photographic film or plate is sensitized for light of certain wavelengths by the addition of various dyes, which enables light passing through the lens to react with the silver halides and form a latent image.

Colour temperature. The colour temperature T_v of a light source is that absolute temperature of a black body at which the black body's emitted radiation has the same spectral energy distribution as the light source. The temperature is expressed in degrees Kelvin ($°K$; $0 °K = -273 \cdot 2 °C$). Colour temperature can also be expressed in mired (1 mired = $10^6/T_v$). Multiplication of mired by $2 \cdot 89$ gives the position of the centre of radiation of the light source expressed in nanometres; e.g. at a colour temperature of 190 mired, the centre of radiation will be $\lambda = 550$ nm.

Complementary colours. A colour is the complementary colour of another when both added together give white light. If, for instance, the blue-green light is isolated from a white continuous spectrum, the remainder of the spectrum will give red light (the complementary colour of blue-green). Complementary colours are:

Red	— Blue-green	Yellow-green	— Violet
Orange	— Light blue	Green	— Purple
Yellow	— Ultramarine blue		

Continuous spectrum. All wavelengths of the light in the emission spectrum of a thermal radiator are represented without gaps. The spectrum produced by a prism is a coherent (continuous) band, without any abrupt discontinuity in intensity.

Dichroism. Some crystals possess the property of changing their absorption power and their colour (illumination with white light), according to the direction of the incident polarized light.

Dispersion. The separating of polychromatic light into its various wavelengths, caused by refraction. The dispersion curve indicates the relationship between the refractive index of a substance and the wavelength of the light.

Efficiency of a light source. The ratio of the total luminous flux emitted by a light source to the energy consumed; it is expressed in lumens per watt.

Emmetropia. The normal condition of eye vision, when parallel rays of light are focused on the retina, with the eye in its relaxed position, and without the help of spectacles.

Focal interception. The distances of an object, or its image, from the poles of the image-forming lens or lens system.

Focal plane shutter. Shutter with a normally variable slit which moves across the back of the camera, as close as possible to the film in the focal plane of the lens.

Lens shutter. Generic term for before-the-lens, between-the-lens and behind-the-lens shutters

Line Spectrum. A discontinuous spectrum consisting of a number of sharp peaks or lines with intervening gaps without any light emission.

Luminance. Measure of the surface brightness of a light source. Since the luminance of different portions of the radiant surface can have different values, a light source is characterized by its mean luminance. The unit of luminance is the nit (one candela per square metre).

Luminous intensity. Luminous intensity is characterized by the luminous flux incident on the surface unit. The unit is the lux. Mean daylight gives an illumination intensity of approximately 3,000 lux. The illumination intensity of a work bench should be 25—30 lux (see also Table 13).

Parallax. When objects lying in different planes are observed in succession from laterally different positions, the more distant objects appear to change their position less than objects close to the observer. Parallax is used in photomicrography as a test to ascertain that image and cross wires are situated in the same plane.

Photo-conductive cell. A photo-cell which has the property of varying its electric conductivity under the influence of light (cadmium sulphide, CdS).

Photo-electric cell. A photo-cell which has the property of generating an electric potential under the influence of light (selenium).

Refraction law. Snell's law states that when a ray of light passes from one medium to another, the sine of the angle of incidence α bears a constant ratio to the sine of the angle of refraction β. Thus

$$n \cdot \sin\alpha = n^1 \cdot \sin\beta$$

where n is the refractive index of the first, n^1 that of the second medium.

Refractive index. The refractive index of a medium for a given wavelength is the ratio of the velocity of light in vacuo to that in the medium ($n = C_v/C_m$).

Transmission factor. The ratio of the total luminous flux transmitted by a body to the incident flux.

LITERATURE REFERENCES

[1] MARQUART, F., The importance of colour temperature for colour photography and how to measure it, *Bild u. Ton* **10**, 241 (1957).

[2] HELBIG, E., Light sources with high luminous emittance, *Mschr. Feinmech. Opt.*, **71**, 5 (1954).

[3] KÖHLER, A., Requirements of light sources for photomicrography and microprojection, *Photogr. Ind., Berlin*, No. 50 (1924).

[4] MARQUART, F., "Werralux" colour test attachment and "A – Z" filters, *Fotografie, Halle*, **12**, 315 (1958).

[5] GERSING, R., Self-igniting arc lamp for photomicrography of constant luminous intensity, *Z. wiss. Mikr.*, **63**, 257 (1957).

[6] KOPEC, U., Modern xenon lamps for colour photography and projection, *Bild u. Ton*, 8, 213 (1955).

[7] IHLN, A., A new xenon lamp for colour sampling, *Bild u. Ton*, **11**, 38 (1958).

[8] BERGNER, J., Movement unsharpness in photomicrography with the miniature camera, *Fototechn. Rdsch. Wiss. Prax.*, **1**, 21 (1956).

[9] WENZEL, F., *Agfa filters*, Fotokinoverlag (Halle) 1957.

[10] *Filter sets for microscopy and photomicrography*, Brochure 30-328-1, Carl Zeiss, Jena.

[11] REINERT, G. G., Use of filters in photomicrography, *Z. wiss. Mikr.*, **51**, 253 (1934).

[12] RÖTGER, A., The Jena interference filter spectroscope, *Explle Tech. Phys.*, **3**, 88 (1955).

[13] KOROLEW, F. A., and Klementjewa, A. J., Dispersion filters of high monochromatosity, *Explle Tech. Phys.* **3**, 44 (1955).

[14] BARNES, R. B., and BORNER, L. G., Christiansen filter effect in the infra-red region, *Phys. Rev.*, **49**, 732 (1936).

[15] MEYER-ARENDT, J., Critical evaluation of UV filters, *Mikroskopie*, **7**, 396 (1952).

[16] OTTO, L., *Das Mikroskop*, p. 120, Uraniaverlag (Leipzig - Jena) 1957.

[17] ANDRES, E., *Preparation method in metallography*, brochure 30-S098-1, Carl Zeiss, Jena.

[18] ROSENBUSCH and WÜLFING, Mikroskopische Physiographie der petrographisch wichtigen Mineralien, p. 4, Schweizerbarth (Stuttgart) 1921.

[19] CLAUSNITZER, H., Mechanical preparation of polished sections, *Bergakademie*, 77 (1957).

[20] KENNEDY, G. C., The preparation of polished thin sections, *Econ. Geol.*, **40**, 353 (1945).

[21] GRABNER, A., The connection between objective, eyepiece and eye, *Mikroskopie*, **10**, 83 (1955).

[22] SCHARF, J.-H., Photographic materials for scientific photomicrography, *Mikroskopie*, **6**, 383 (1951).

[23] BERGNER, J., Faithful object rendering in photomicrography with the miniature camera, *Fotografie, Halle* **10**, 184 (1956).

[24] BRÜNNER, G., 35 mm document film and photomicrography, *Mikrokosmos*, **45**, 188 (1956).

[25] NÜRNBERG, A., *Agfa-Photo-Materialien für Wissenschaft und Technik*, VEB Knapp-Verlag (Halle) 1954.

[26] KRAUSS, B., Monochrome reversal film for test exposures for colour photography, *Photogr. u. Wiss.*, **6**, 12 (1957).

[27] STECHE, W., Exposure in photomicrography, *Mikrokosmos*, **43**, 191 (1953/4).

[28] WIELAND, M., A method for exposure determination in photomicrography, *Z. wiss. Mikr.*, **53**, 183 (1936).

[29] HASELMANN, H., A new light adjustment device for the microscope, *Mikroskopie*, 8, 328 (1953).

[30] HAUSER, F., Simultaneous incident and transmitted light illumination, *Zeiss Nachr.*, **1**, 30 (1932).

[31] SCHARDIN, H., Schlieren methods, *Ergebn. exakt. Naturw. XX*, Berlin (1942).

[32] MEYER-ARENDT, J., Schlieren methods for transmitted and incident light, *Mikroskopie*, **12**, 21 (1957).

[33] TOEPLER, A., Observations with the schlieren method, *Ostwalds Klass.* No. 157-8.

[34] KÖHLER, A., and LOOS, W., Phase contrast, *Naturwissenschaften*, **29**, 49 (1941).

[35] BEYER, H., Shape of aperture diaphragm and phase contrast method, *Jena Jb.*, 1953, p. 162.

[36] BURRI, C., *Das Polarisationsmikroskop*, Verlag Birkhäuser (Basel) 1950.

[37] GAUSE, H., The polarized light microscope, *Jena Nachr.* 8, 134 (1959).

[38] BERGNER, J., Polarized light microscope and colour photography, *Bild u. Ton*, 9, 66 (1956).

[39] SHORT, M. N., Microscopical determination of the ore minerals, *Bull. U. S. Geol. Surv.* 914 (1940).

[40] BERGNER, J., The attachment camera for polarized light photomicrography, *Jena Rdsch.*, **4**, 62 (1959).

[41] SCHARF, J.-H., Polarized light photomicrography and macrophotography, *Naturwiss. Rdsch.*, **7**, 431 (1954).

[42] RÖSCH, S., Measurement of colour rendering characteristics of miniature colour films, *Photogr. Korr.*, **95**, 7 (1959).

[43] REUMUTH, H., and KÖHLER, W., Infra-red photography in textile chemistry and testing, *Z. ges. Textilind.*, No. 22 (1936).

[44] KRAFT, P., Infra-red photomicrography, *Z. dtsch. geol. Ges.*, 84, 651 (1932).

[45] COULON, F., Infra-red microscopy, *Mikrokosmos*, **43**, 284 (1953/4).

[46] MALMQVIST, D., Infra-red axial images of opaque minerals, *Zbl. Miner. Geol. Paläont.* A, 209 (1935).

[47] TERENIN, A., A photographic method in infra-red, *Z. Phys.*, **23**, 294 (1924).

[48] WEBER, K., Infra-red photography with the sensitized Herschel effect, *Photogr. Korr.*, **75**, 22 (1939).

[49] GÖRLICH, P., Using light-sensitive organs in optical microscopy, *Wiss. Ann.*, **5**, 724 (1956).

[50] LÜHR, F., and NÜRNBERG, A., *Agfa-Rezepte*, VEB Filmfabrik Agfa (Wolfen) 1951.

[51] REINERT, G. G., Long-wave photomicrography, *Z. wiss. Mikr.*, **50**, 344 (1933).

[52] GÖRLICH, P., *et al.*, Series-produced image converters, *Z. angew. Phys.*, **9**, 561 (1957).

[53] KÖHLER, A., New methods in UV photomicrography, *Naturwissenschaften*, **21**, 165 (1933).

[54] TWIDLE, G. G., The spectral reflectivity of aluminized mirrors, *Brit. J. Appl. Phys.*, **8**, 337 (1957).

[55] HÖRMANN, H., and SCHOPPER, E., Photographische Schichten für die wissenschaftliche Photographie, *Veröff. Agfa* VII, 108 (1940).

[56] OTTO, L., Fluorescence photomicrography, *Fotografie, Halle*, **7**, 158 (1953).

[57] BRUMBERG, E. M., and KRYLOVA, T. N., Interference beam-splitter in fluorescence microscopy, *Z. angew. Biol. (USSR)*, 461 (1953).

[58] DANCKWORTT, P. W., *Lumineszenzanalyse im filtrierten ultravioletten Licht*, Akad. Verlagsges. Geest & Portig (Leipzig) 1949.

[59] STRUGGER, S., *Fluoreszenzmikroskopie un Mikrobiologie*, Schaper-Verlag, (Hannover) 1949.

[60] PRESTING, W., and DORNICK, H., Fluorescence photomicrography and organic chemistry, *Fototechn. Rdsch. Wiss. Prax.*, **1**, 36 (1956).

[61] BERGNER, J., Electronic flash and photomicrography, *Feingerätetechnik*, **1**, 394 (1952).

[62] MICHEL, K., Electronic flash and photomicrography, *Photogr. u. Forsch.*, **5**, 140 (1952).

[63] STACH, E., Incident light objectives with oil immersion hoods, *Mikroskopie*, **12**, 232 (1957).

[64] HAUSMANN, G., Photography of a comparison scale next to the specimen by means of the multiple exposure attachment, *Zeiss Nachr.*, 4, 155 (1943).

[65] KROLL, P., Under-exposure and prolonged development, *Fotografie, Halle*, **6**, 391 (1952).

[66] OTTO, L., Under-exposure and prolonged development, *Fotografie, Halle*, **7**, 75 (1953).

[67] GUNTERMANN, S. ,Tolerance in colour temperature with colour photography, *Bild u. Ton*, **12**, 170 (1959).

[68] ROSS, K., Photography by UV and infra-red illumination, *Zeiss Nachr.* 1, 19 (1932).

TABLES

TABLE 1

SUMMARY OF REPRODUCTION RATIOS WITH POSITIVE LENSES

Object y		Object y'			Ratio $y':y$
position	type	position	type	image	
$-\infty - 2\bar{f}$	real	$F_2 - 2f'$	real	inversed	$0:1-1:1$
$2\bar{f} - F_1$	real	$2f' - +\infty$	real	inversed	$1:1-\infty:1$
$F_1 - N_1$	real	$-\infty - N_2$	virtual	upright	$\infty:1-1:1$
$N_1 - +\infty$	virtual	$N_2 - F_2$	real	upright	$1:1-0:1$

TABLE 2

SUMMARY OF REPRODUCTION RATIOS WITH NEGATIVE LENSES

Object y		Object y'			Ratio $y':y$
position	type	position	type	image	
$-\infty - N_1$	real	$F_2 - N_2$	virtual	upright	$0:1 - 1:1$
$N_1 - F_1$	virtual	$N_2 - +\infty$	real	upright	$1:1 - \infty:1$
$F_1 - 2\bar{f}$	virtual	$-\infty - 2f'$	virtual	inversed	$\infty:1 - 1:1$
$2\bar{f} - +\infty$	virtual	$2f' - F_2$	virtual	inversed	$1:1 - 0:1$

DEPTH OF FIELD t_1 IN MACROPHOTOGRAPHY FOR STANDARD SCALES OF REPRODUCTION, AND OPTIMUM DIAPHRAGM OF THE ENTRANCE PUPIL WITH CONSIDERATION OF THE FOCAL LENGTHS OF THE OBJECTIVES "M". TOLERANCE FOR ENTRANCE PUPIL DIAMETER: 0.1 mm. CALCULATIONS ACCORDING TO EQUATIONS (31a) AND (31b). CIRCLE OF CONFUSION: $^1/_7$ mm

$M:1$	t_1 mm	Diameter of entrance pupil in millimetres							
		$f=120\,mm$	$f=90\,mm$	$f=60\,mm$	$f=45\,mm$	$f=30\,mm$	$f=20\,mm$	$f=15\,mm$	$f=10\,mm$
1	71·4	1							
1·25	45·7	1·1							
1·6	28·0	1·3	1						
2	17·9	1·4	1·1						
2·5	11·4	1·7	1·3						
3·2	7·0	2·0	1·5	1					
4	4·5	2·4	1·8	1·2					
5	2·9	2·8	2·2	1·4	1·1				
6·3	1·8	*3·5*	2·6	1·7	1·3				
8	1·1	4·3	3·2	2·2	1·6	1·1			
10	0·72	5·2	*3·9*	2·6	2·0	1·3			
12·5	0·46	6·5	4·9	3·2	2·4	1·6	1·1		
16	0·28	8·2	6·1	*4·1*	3·1	2·0	1·4	1	
20	0·18	10·1	7·6	5·1	*3·8*	2·5	1·7	1·3	
25	0·11	12·5	9·4	6·2	4·7	*3·1*	2·1	1·6	
32	0·07			8·0	6·0	4·0	2·7	2·0	1·3
40	0·04			9·9	7·4	4·9	*3·3*	2·5	1·6
50	0·03			12·3	9·2	6·1	4·1	3·1	2·0
63	0·02						5·1	*3·8*	2·6
80	0·01							4·9	*3·2*
100	0·007							6·1	4·0

The cursive values are the limits for a camera extension of 1 metre

TABLE 3 b

DEPTH OF FIELD t_2 IN μ FOR PHOTOGRAPHY WITH THE COMPOUND MICROSCOPE, FOR STANDARD SCALES OF REPRODUCTION IN THE USEFUL RANGE. CALCULATED ACCORDING TO EQUATION (32) FOR JENA TRANSMISSION OBJECTIVES. CIRCLE OF CONFUSION: $^1/_7$ mm

M:1	Achromatic and apochromatic objectives						
	8 0·20	10 0·30	20 0·40	20 (40) 0·65	40 0·95	HI 90 1·30	HI 60 1·40
100	7·14						
125	5·72						
160	4·46	2·98					
200	3·57	2·38	1·79				
250		1·91	1·43				
320		1·49	1·12	0·69			
400			0·89	0·55			
500				0·44	0·30		
630				0·35	0·24	0·27	0·25
800					0·19	0·21	0·19
1000					0·15	0·17	0·16
1250						0·13	0·12

M:1	Plano-objectives					
	2.5 0·07	4 0·11	6.3 0·16	16 0·32	40 0·65	HI 100 1·25
32	63·8					
40	51·0					
50	40·8	26·0				
63	32·4	20·6				
80		16·2	11·15			
100		13·0	8·93			
125			7·14			
160			5·58	2·79		
200				2·23		
250				1·79		
320				1·40	0·69	
400					0·55	
500					0·44	
630					0·35	0·28
800						0·22
1000						0·17
1250						0·14

TABLE 4

LIGHT SOURCES FOR PHOTOMICROGRAPHY

Type	Volt	Amp	Watts	Mean luminance (cd/cm²)	Radiant surface (mm²)	Colour temperature (°K)	Relative running cost*
projection lamp 6 v, 15 w	6	2·5	15	750	1·4×1·4	2,850	1·2
projection lamp 6 v, 15 w flat coil	6	2·5	15	900	2·0×1·8	2,990	1
projection lamp 6 v, 30 w	6	5	30	1,000	2·2×2·0	2,950	2·8
projection lamp 12 v, 100 w	12	8·3	100	2100	3·5×3·5	3,100	1·6
projection lamp 12 v, 100 w flat coil	12	8·3	100	2850	4·2×2·6	3350	5·2
carbon arc lamp	50 DC	6	300	17,000	2·5 dia	3,850	5·6
do.	50 AC	10	500	11,000	2·5 dia	3,450	5·6
xenon high-pressure lamp XBO 100	16—20	7—5	100	10,000	0·7×1·5	5,800**	18·8
xenon high-pressure lamp XBO 200 F	20—25	10—8	200	6,000	1·5×2·5	5,200	12
mercury high-pressure lamp HBO 50	35—50	1·7—1·2	50	20,000	0·6×1·2	—	16
mercury high-pressure lamp HBO 200	55—75	4·0—2·7	200	20,000	1·4×2·5	—	16

* relative approximate values, referring to useful life.

** when the bright cathode spot is not used for illumination, the colour temperature is reduced to approx. 5,200 °K.

TABLE 5

HOMAL EYEPIECES

Name	Field number (mm)	Focal length (mm)	To be used with apochromats:	
Homal II	15	—70	15×/0·30	∞/0
			60×/0·95	∞/0
			10 /0·30	160/0·17
Homal IV	14	—37·5	15×/0·30	∞/0
			32×/0·65	∞/0
			60×/0·95	∞/0
			90×/1·30	∞/0
			20 /0·65	160/0·17
Homal VI	8	—20	90×/1·30	∞/0
			40 /0·95	160/0·17
			60 /1·00	160/0·17
			60 /1·40	160/0·17
			90 /1·30	160/0·17

TABLE 6

ADAPTATION OF THE HOMAL II EYEPIECE TO THE CAMERA EXTENSION

Camera extension (mm)	Division to be adjusted	Total size of reproduction with apochromat 10/0·30
200	5	50:1
265	6	60:1
340	7	70:1
410	8	80:1
490	9	90:1
560	10	100:1
660	12	120:1

TABLE 7

COMPARISON OF EXPOSURE TIMES REQUIRED TO OBTAIN IDENTICAL
DENSITY WITH CAMERA SHUTTERS AND TIMERS

Illumination in film plane (lux)	Camera shutter (seconds)	Timer (seconds)
25·6	0·07	0·3
12·8	0·15	0·4
6·4	0·30	0·5
3·2	0·60	0·8
1·6	1·2	1·3
0·8	2·5	2·5
0·4	5·0	5·0

TABLE 8a

FILTER FACTORS FOR PHOTOMICROGRAPHY (COLOUR TEMPERATURE 2,850°K)

Filter	Thickness (mm)	Orthochromatic	Panchromatic
BG 7	3	8	12
BG 12	2	10	20
BG 17	4	1·5	1·5
BG 33	2	2·5	3
VG 4	2	3	1·5
VG 8	2	3·5	4
VG 9	4	16	18
GG 11	2	2	1
OG 3	2	—	2·5
RG 2	2	—	140
NG 5	2	4	4
NG 5	4	8	8
BG 7 + GG 11	3 + 2	20	20

FILTER FACTORS FOR INTERFERENCE FILTERS, RELATED TO ORWO-FILMS
(APPROXIMATED VALUES)

λ_{max} (nm)	NP 18	NO 20	UK 14	UT 16
486 (F)	900	25	200	150
551	175	60	30	60
589 (D)	400	—	600	250
656 (C)	15,000	—	150	600

TABLE 9

SCALE OF REPRODUCTION AND ACTUAL RESOLUTION

M	y^*	M	y^*
1·0 :1	0·32	40:1	0·0080
1·25:1	0·25	50:1	0·0063
1·6 :1	0·20	63:1	0·0050
2·0 :1	0·16	80:1	0·0040
2·5 :1	0·12	100:1	0·0032
3·2 :1	0·10	125:1	0·0025
4·0 :1	0·080	160:1	0·0020
5·0 :1	0·063	200:1	0·0016
6·3 :1	0·050	250:1	0·0012
8·0 :1	0·040	320:1	0·0010
10·0 :1	0·032	400:1	0·00080
12·5 :1	0·025	500:1	0·00063
16 :1	0·020	630:1	0·00050
20 :1	0·016	800:1	0·00040
25 :1	0·012	1,000:1	0·00032
32 :1	0·010	1,250:1	0·00025

The scales of reproduction are graded in accordance with the R 10 standard range.

$$* \; y = \frac{0·32}{M} \; mm.$$

TABLE 10

NEGATIVE MATERIALS FOR PHOTOMICROGRAPHY

	Type	Colour sensit- ivity	Speed	Developer		Resolution (lines/mm)
Plates	Agfa Mikro	ortho	20 ASA	Final	8 min.	85
	Agfa Isochrom F	ortho	50 ASA	Final	8—10 min.	65
	Agfa Isopan F	pan	40 ASA	Final	8—10 min.	67
Miniature films	Agfa Isochrom F	ortho	40 ASA	Atomal F	7—8 min.	70
	Agfa Isopan FF	pan	8 ASA	Atomal F	4 min.	100
	Agfa Isopan F	pan	40 ASA	Atomal F	7—8 min.	75

TABLE 11

EXPOSURE TIMES FOR TEST EXPOSURES
WITH THE PLATE HOLDER SLIDE

1st	2nd	3rd	4th	5th	6th	7th	Strip
1	1	1	1	1	1	1	
	1	1	1	1	1	1	
		2	2	2	2	2	individual
			4	4	4	4	exposures
				8	8	8	
					16	16	
						32	
1	2	4	8	16	32	64	total time

TABLE 12

ILLUMINATION IN MINIATURE PHOTOMICROGRAPHY

Transmitted light	Minimum (lux)	Maximum (lux)	Main values (lux)
bright field	0·4	44	8 —13
phase contrast	0·2	3·5	0·5— 1·3
fluorescence	0·08	2	0·2
polarization	0·02	44	2 — 4

TABLE 13

CALIBRATION EXAMPLE OF AN EXPOSURE METER

Strip	Galvanometer deviation	Exposure time t	Density D	$t_{D=1}$
1	1,000	0·2 s	0·49	0.9 s
		0·5 s	0·81	
		1 s	1·04	
		2 s	1·30	
2	500	0·5 s	0·61	2.1 s
		1 s	0·78	
		2 s	0·98	
		4 s	1·23	
3	250	1 s	0·57	4.8 s
		2 s	0·74	
		4 s	0·95	
		8 s	1·18	
4	125	4 s	0·70	8.6 s
		8 s	0·97	
		15 s	1·18	
		30 s	1·44	
5	62	8 s	0·72	18 s
		15 s	0·93	
		30 s	1·24	
		60 s	1·46	
6	31	15 s	0·66	40 s
		30 s	0·88	
		60 s	1·16	
		2 min	1·40	
7	16	30 s	0·58	84 s
		60 s	0·85	
		2 min	1·11	
		4 min	1·35	
8	8	1 min	0·60	220 s
		2 min	0·78	
		4 min	1·04	
		8 min	1·34	
9	4	4 min	0·65	12 min
		8 min	0·87	
		15 min	1·10	
		30 min	1·40	

222

TABLE 14

SCALES OF REPRODUCTION AND OBJECT FIELD VALUES
WITH MACROPHOTOGRAPHIC-OBJECTIVES (I)

Objective f_{obj}	Miniature camera		Bellows camera 9×12 cm	
	scale of reproduction	encompassed object field (mm)	scale of reproduction	encompassed object field (mm)
10 mm	12·5:1	2·0× 3.0	25 :1—63 :1	3·3× 4.5—1·3×1·8
20 mm	5 :1	4·8× 7.2	12·5:1—25 :1	6·6× 9 —3·3×4·5
30 mm	3·2:1	7·5×11.2	8 :1—16 :1	10·5×14.2—5·2×7·1
45 mm	2 :1	12 ×18	5 :1—12·5:1	16·8×22.8—6·6×9·0

These values are approximate and depend on the arrangement. The object field's values can be increased by the use of extension tubes (miniature cameras) or a mirror attachment (bellows cameras).

TABLE 15

SCALES OF REPRODUCTION AND OBJECT FIELD VALUES
WITH MACROPHOTOGRAPHIC-OBJECTIVES (II)

Objective f_{obj}	Scale of reproduction	Encompassed object field (mm)
60 mm	3·2:1—6·3:1	26×36—13×18
90 mm	1·6:1—3·2:1	52×72—26×36
120 mm	1 :1—2·5:1	dia 82—33×45
135 mm	1 :1—1·6:1	dia 82—52×72

These values refer to the scales of reproduction obtained with a 9×12 cm bellows camera and have been approximated to the nearest standard number.

TABLE 16

FILTERS FOR PHOTOMICROGRAPHY OF ORE MINERALS[39]

Contrast increase between:	Filter needed
chalcopyrite and galena	blue
chalcopyrite and pyrite	blue
chalcopyrite and chalcocite	green
chalcopyrite and bornite	green
chalcocite and covellite	blue
specular iron and magnetite	blue
bornite and covellite	blue
argentite and galena	green
bornite and tennantite	green
zinc blende and other minerals	yellow-green, yellow
gangue and other minerals	yellow-green, yellow

TABLE 17

PARALLAX AND ABSCISSAE DIFFERENCE

Distance of point under observation (metres)	Compared with the direction of vision to infinity	
	parallax at $2\bar{d} = 60$ mm	abscissae difference $\Delta x'$ reference distance $a' = 25$ cm (mm)
∞	0	0
10	20·6′	0·75
5	41·4′	1·5
3	68·8′	2·5
2·5	1°22′	3
1·5	2°18′	5
1	3°26′	7·5
0·75	4°36′	10
0·375	9°10′	20
0·25	13°40′	30

INDEX